T0205423

Studies in Big Data

Volume 62

The series "Studies in Big Data" (SBD) publishes new developments and advances in the various areas of Big Data- quickly and with a high quality. The intent is to cover the theory, research, development, and applications of Big Data, as embedded in the fields of engineering, computer science, physics, economics and life sciences. The books of the series refer to the analysis and understanding of large, complex, and/or distributed data sets generated from recent digital sources coming from sensors or other physical instruments as well as simulations, crowd sourcing, social networks or other internet transactions, such as emails or video click streams and other. The series contains monographs, lecture notes and edited volumes in Big Data spanning the areas of computational intelligence including neural networks, evolutionary computation, soft computing, fuzzy systems, as well as artificial intelligence, data mining, modern statistics and Operations research, as well as self-organizing systems. Of particular value to both the contributors and the readership are the short publication timeframe and the world-wide distribution, which enable both wide and rapid dissemination of research output.

** Indexing: The books of this series are submitted to ISI Web of Science, DBLP, Ulrichs, MathSciNet, Current Mathematical Publications, Mathematical Reviews, Zentralblatt Math: MetaPress and Springerlink.

More information about this series at http://www.springer.com/series/11970

Mario Kubek

Concepts and Methods for a Librarian of the Web

Mario Kubek
FernUniversität in Hagen
Lehrgebiet Kommunikationsnetze
Hagen, Nordrhein-Westfalen, Germany

ISSN 2197-6503 ISSN 2197-6511 (electronic)
Studies in Big Data
ISBN 978-3-030-23138-5 ISBN 978-3-030-23136-1 (eBook)
https://doi.org/10.1007/978-3-030-23136-1

This Springer imprint is published by the registered company Springer Nature Switzerland AG
The registered company address is: Gewerbestrasse 11, 6330 Cham, Switzerland

Preface

If the World Wide Web (WWW, web) is considered a huge library, it would need a capable librarian or at least services that could carry out this person's tasks in a comparable manner and quality. Google and other web search engines are more or less just keyword-addressed databases, and cannot even remotely fulfil a librarian's manifold tasks such as the classification and cataloguing of publications as well as the mediation between library resources and users. Therefore, librarians can provide much better support for long-term and in-depth research tasks. In order to algorithmically and technically address these problems, this book introduces the 'Librarian of the Web', a novel concept, which comprises librarian-inspired approaches, methods and technical solutions for decentralised text search in the WWW using peer-to-peer (P2P) technology. To this end, the following aspects are highlighted:

1. A graph-based approach is introduced to semantically represent both search queries and text documents of arbitrary length alike.
2. For the needed library construction and management, novel hierarchical clustering and classification methods that make use of this approach are presented and evaluated. It is shown that these methods work properly even when the document base is growing.
3. New measures are suggested to locally evaluate the quality of user queries before they are sent out. These measures can be used to give users instant feedback on whether to reformulate a query or not. In this regard, the well-known and widely used measures recall and precision in the field of information retrieval are critically discussed.
4. Novel methods to route these queries and their representations within the constructed library to peers with potentially matching documents are presented.
5. The first implementation of this concept, the P2P-client called 'WebEngine', is elaborated on in detail. Its general architecture, software components, working principles and core algorithms are highlighted.

The structure of the generated P2P-network is directly induced by exploiting the web's explicit topology (links in web documents). Thus, a decentralised web search system is created that—for the first time—combines state-of-the-art text analysis techniques with novel, effective and efficient search functions as well as methods for the semantically oriented P2P-network construction and management.

Hagen, Germany Mario Kubek
April 2019

Contents

Chapter 1
Introduction

1.1 Motivation

Public libraries are often lonesome places these days, because most of the information, knowledge and literature is made available in the omnipresent Internet, especially in the World Wide Web (WWW, web) as a major part of it. It seems that the times are forgotten, when librarians collected giant amounts of books, made them refindable by huge catalogue boxes containing thousands of small cards and archived them using their (own) special scheme in the right place in many floors consisting of a maze of shelves. In addition to these tasks and loaning books, they had time to support library users by giving them advises on where to find the wanted information quickly and maybe to even tell them the latest news and trends as well.

Currently, there is no information system available that can even remotely fulfil all of these tasks in an acceptable manner. However, as the amount of available textual data (especially in the WWW) is steadily growing and the data traffic volume for private web usage and sending e-mails reached 9170 PetaByte per month in 2016 [134], there is an urgent need for such a system which is—similar to the human librarian in her/his role as a caregiver—to be regarded as an active technical intermediary between users and resources such as textual documents. A respectively designed information system has to be able to autonomously

- provision, archive and manage information in many formats,
- provide an efficient information access (e.g. offer topic suggestions and topical 'signposts' as well as generate bibliographies),
- proactively procure information on the basis of identified information needs (demonstrate information literacy i.e. 'the ability to know when there is a need for information as well as to identify, locate, evaluate, and effectively use that information for the issue or problem at hand' and become a pathfinder for possibly relevant information) and
- carry out search tasks while filtering out unimportant or even unsolicited information (one could simply speak of data in this case, too) by applying classification methods.

M. Kubek, *Concepts and Methods for a Librarian of the Web*,
Studies in Big Data 62, https://doi.org/10.1007/978-3-030-23136-1_1

This implicitly means that the system must be able to perform search tasks on its own and on behalf of the user when requested. A system offering these features would especially facilitate in-depth research in a sustainable manner which strongly differs from short-term or adhoc search tasks. Particularly, in-depth research

- is an iterative and interactive process,
- has a context and history,
- consists of different search paths and directions,
- means to learn from positive and negative feedback and
- influences the objects being searched for as well (as an example: the most requested objects by experts in a field would likely be of relevance given respectively categorised queries, would therefore be returned first and would more easily become subjects to further scientific investigations).

These points imply that the mentioned information system is able to cope with and learn from dynamically changing contexts such as (even short-term) shifts in information needs and topical changes in the local document base as well as to identify possibly upcoming new trends or new concepts from various information streams. When the system takes into account the history of past and ongoing search processes in form of search paths consisting of queries and result sets, a navigation in previous search steps is possible. By this means and based on the learned concepts and their relationships, the system can also interactively provide alternative search directions as well as topic suggestions to follow.

While these functionalities are particularly of benefit for users conducting research, they address the problem of refinding information which is aggravated by the so-called 'Google effect', sometimes referred to as 'digital amnesia' [115], too. The main findings related to this effect are that people tend to forget information when they assume that it can be found again using digital technology and that they are more likely to remember how they previously found a certain information using search engines (the search path) instead of the information itself. This indicates that people are generally satisfied with the obtained search results (the relevant ones are presented first); otherwise the feeling would arise that it could be hard to find them again at a later time and it could be better to memorise the respective information. As this development—at least to a certain extent—affects the way research is carried out today [123], the mentioned system's functionalities of storing, retrieving and suggesting search paths should also facilitate the recovery of previous search processes and their results.

1.2 A Critical Look at Current Web Search Engines

It is definitely a great merit of the WWW to make the world's largest collection of documents of any kind in digital form easily available at any time and any place without respect to the number of copies needed. It can therefore be considered to be the knowledge base or library of mankind in the age of information technology.

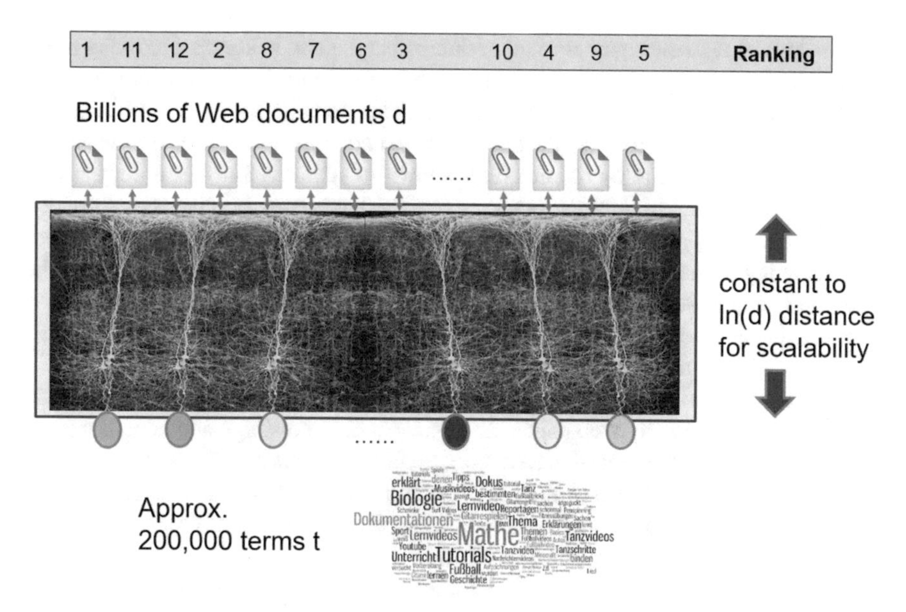

Fig. 1.1 The dimensionality problem of the WWW

Google (https://www.google.com/), as the world's largest and most popular web search engine with its main role to connect information and the place/address where it can be found, might be the most effective, currently available information manager.

Even so, in the author's opinion, Google and Co. are just the mechanistic, brute force answer to the problem of effectively managing the complexity of the WWW and handling its big data volumes. As already discussed e.g. in [31], a copy of the web is established by crawling it and indexing web content in big reverse index files containing for each occurring word a list of files in which they appear. Complex algorithms try to find those documents that contain all words of a given query and closely related ones. Since (simply chosen) keywords/query terms appear in millions of (potentially) matching documents, a relevance ranking mechanism must avoid that all of these documents are touched and presented in advance to the user (see Fig. 1.1).

In the ranking process, the content quality and relative position of a document in the web graph as well as the graph's linking structure are taken into account as important factors. In addition to organic search results obtained from this process, advertisments are often presented next to them which are related to the current query or are derived from personalisation efforts and detected user's interests. In both cases, web search engines do not take into account probably existing user knowledge. To a certain extent, this procedure follows a top-down approach as this filtering is applied on the complete index for each incoming query in order to return a ranked list of links to matching documents. The top-ranked documents in this list are generally useful.

However, due to the sheer amount of data to handle and in contrast to the bottom-up approach of using a library catalogue or asking a librarian or human expert for guidance in order to find (more) actually relevant documents, the search engines' approach is less likely to return useful (links to) documents at an instant when the search subject's terminology is not fully known in advance. Furthermore, the web search results are not topically grouped, a service that is usually inherently provided by a library catalogue. Therefore, conducting research using web search engines means having to manually inspect and evaluate the returned results, even though the presented content snippets provide a first indication of their relevance.

From the technical point of view, the search engine's architecture carrying out the mentioned procedures has some disadvantages, too:

On the one hand, in order to be able to present recent results, the crawlers must frequently download any web pages. To achieve a high coverage and actuality (web results should cover contents that have been updated in the last 24 h with a probability of at least 80%), they cause avoidable network load. Problems get bigger, once the hidden web (deep web) is considered besides regularly accessible HTML (Hypertext Markup Language) pages (surface web), too.

On the other hand, modern web topology models (like the evolving web graph model [16]) emanate from the fact that there are linear as well as exponential growth components, if the overall number of websites is considered. From the point of view of complexity theory, it becomes clear that there might be no memory space available to keep and archive all versions of (previously) existing and newly emerging sites in the future.

1.3 Main Contributions

As the main contribution of this book, a basic, new concept for decentralised web search, subsumed under the name 'Librarian of the Web', that is derived from the foregoing considerations and identified shortcomings of current web search engines shall be derived and explained in the subsequent chapters. To this end, a number of subordinate concepts, supporting methods and their evaluations will be presented. These methods are characterised by their local working principles, making it possible to employ them on diverse hardware and to take into account user-specific knowledge to carry out and improve search tasks. The implementation in form of an interactive, librarian-and physics-inspired peer-to-peer (P2P) web search system, called 'WebEngine', will be elaborated on in detail, too.

The client software of this system will at its core comprise of several components responsible for the storage, retrieval and semantic analysis of text documents, for the P2P-network construction and maintenance as well as for the execution of local and network-wide search tasks. Thus, a decentralised web search system is created and formed that—for the first time—combines modern text analysis techniques with novel and efficient search functions and a semantically induced P2P-network construction and management. In a more abstract and general view, the system will therefore make use of analysis (text mining and query interpretation) and synthesis (library and network construction) methods, whereby the latter ones depend on the former ones.

It is not the author's intention to replace existing web search engines per se, as their usefulness is uncontested. However, an alternative, integrated approach to web search under the motto 'The Web is its own search engine.' shall be presented, which was conceived to inherently support users with particular search and research tasks. It will be shown, that this approach will make it possible for the P2P-network to restructure itself by means of self-organisation such that it becomes maintainable and searchable without any central authority.

1.4 Outline

This system's design is reflected in the threepart organisation of this book. The first five chapters provide an introduction to the concepts and algorithms needed for the understanding of the main contributions, which are in turn discussed in detail in the Chaps. 6 to 9. The concluding Chap. 10 provides a view on the future of decentralised web search and summarises the major findings and realisations presented before. In greater detail, the book is structured as follows:

- The following Chap. 2 reviews the many services provided by libraries with a focus on the activities of librarians. It is further analysed which of those activities can and must be technically realised in decentralised web search systems. Therefore, this chapter acts as a topical anchor and reference for the remainder of the book.
- In Chap. 3, an introduction to the architecture and working principles of current web search engines is given. Also, approaches for interactive and semantic web search are discussed. After providing a critical review of these solutions, the need for the new web search concept of the 'Librarian of the Web' is motivated. For this purpose, preliminary works, which have influenced its conception, are described as well.
- Since the 'WebEngine', the first implementation of the concept of the 'Librarian of the Web', has to mainly deal with natural language text to carry out its tasks, Chap. 4 is dedicated to basic and advanced methods for natural language processing and text mining. Here, the focus is set on approaches to determine characteristic terms or words in texts and to measure their semantic relatedness. Furthermore, algorithms for the grouping of words and texts are discussed.
- Chapter 5 presents various approaches, tools and services developed by the author to support the local and context-based search for web documents. Due to their design and working principles, they are perfectly suited for their future integration in the 'WebEngine' and are thus of relevance to be presented in this book.
- As the first part of the main contributions, Chap. 6 introduces a new graph-based concept for representing texts by single, descriptive terms, called centroid terms. This concept is of utmost importance in order to understand the way of how

text documents and search queries are being analysed, represented, organised and searched for by the 'WebEngine'.

- Chapter 7 extensively covers the newly developed, top-down algorithms for the decentralised library generation and management, the graph-based, hierarchical document clustering and the generation of child nodes (peers). Besides their formal description, the quality of the generated document clusters is evaluated in a number of experiments. A general discussion on the obtained results concludes this chapter.
- Based on the findings in the previous two chapters, Chap. 8 shows that centroid terms are actually helpful in supporting and improving interactive search processes in the hierarchical document structures. Also, two new graph-based measures to qualitatively evaluate queries in interactive search sessions are presented. These measures immediately indicate whether a user should reformulate a query. Furthermore, a method for user-based document ranking is introduced.
- Chapter 9 describes the 'WebEngine', the first P2P-based implementation of the 'Librarian of the Web'. The client's architecture and software components are elaborated on in detail. Also, its working principles and algorithms for the management of the peer neighbourhood and the bottom-up library construction and maintenance are highlighted. Also, the utilisation of graph database systems, Neo4j (https://neo4j.com) in particular, is explained.
- Chapter 10 summarises the main contributions of this book and provides an outlook on possible extensions of the presented concept of the 'Librarian of the Web' and its implementation.

As the chapters build upon each other, it is advised to read them in the order presented. This way, a deep understanding of all discussed approaches, methods and solutions is reached.

Chapter 2
Library Services

2.1 The Tasks of a Librarian

A library is popularly considered a collection of books or a building in which they are stored and cared for. While this understanding is generally correct, it is—at the same time—somewhat limited. There are a large number of common services that libraries provide. Here, the foremost function of libraries is to supply the public and institutions with information [36]. In order to be able to do so, they collect, catalogue and make published literature available in form of various types of media or resources such as books, magazines, newspapers and digital storage media like CDs and DVDs. The access to literature libraries make available is open, unrestricted and usually provided free of charge or for a reasonable price. Furthermore, the archiving of these resources is another important business of libraries which ensures the continuity of literary works. This step usually comprises additional tasks, especially when preserving book collections. For example, they have to be properly restored (when needed) and specially cared for (e.g. select a dry storage location with steady temperature and humidity levels). Also, their digitisation might be part of an archiving process in order to make contents searchable and easily transferable in electronic form. These tasks are usually carried out by librarians and archivists, depending on their specialisation.

The activities of librarians can be roughly classified into collection-centered and user-centered ones. While the collection-centered activities are related to the management of collected media and comprise the

- selection,
- acquisition,
- processing,
- cataloguing,
- care and
- archiving

M. Kubek, *Concepts and Methods for a Librarian of the Web*,
Studies in Big Data 62, https://doi.org/10.1007/978-3-030-23136-1_2

of media, the user-centered activities include

- providing information,
- giving advice,
- organising and carrying out training courses,
- lending media and
- stocktaking (includes the recording and management of local as well as inter-library book loans).

A librarian's focus can therefore be either on collection-centered or user-centered activities. In the following subsections, the two most important activities of librarians are discussed in detail. They are particularly of relevance to the herein presented algorithms and technical solutions.

2.1.1 *Librarians as Intermediaries*

As information needs are constantly rising, the most important task of librarians is to mediate between requesting patrons and proper literature as well as information that satisfy their information needs. In that regard, they become active intermediaries in a search or research process. Therefore, they must be able to instruct library users properly on how they can find relevant information in the library (location of literature) with respect to the field of the subject at hand.

This role as a 'knowledge mediator' becomes even more important in the digital era [9] since it does not only require a solid educational background and good communication skills but encompasses the ability to deal with information technology and to make use of respective tools for data management and manipulating data, too. This changing role makes it possible for librarians to play an active and greater part in research processes and is thus especially of importance for the profession of the librarian as such in the future. Specialised librarians such as the so-called teaching librarians even give lectures on information literacy or competency, i.e. 'the ability to know when there is a need for information as well as to identify, locate, evaluate, and effectively use that information for the issue or problem at hand' [3].

Summarising, the provision of information and information sources is the most important service offered by librarians from the library user's point of view. For that reason, the quality of this service should be measured [5] by the following five indicators:

- Is the information desk visible and easily approachable?
- Does the librarian show interest in the user's request?
- Does the librarian listen carefully to the user and inquire openly if needed?
- Does the librarian make use of proper information resources and the correct research strategy (covers respective explanations to the user as well)?
- Are follow-up questions asked in order to determine if the user unterstood the information provided?

Thus, the librarian has to be friendly, ready to help, supportive and patient in order to provide good service. At the same time, the librarian must be able to adapt to a user's needs while keeping professional distance and providing equals service to all users (avoiding to outsmart particular users). Also, besides a good general education, language expertise and communication skills, a librarian has to have the ability to think in a structured manner and an interest in modern information technology.

2.1.2 Cataloguing Collections

Besides handling user requests, librarians usually are actively involved in the process of cataloguing media which is needed to keep track of the library's holdings and ultimately to make them refindable and is therefore at the core of the mentioned collection-centered activities. From a historical viewpoint, catalogues in book form, card catalogues and modern Online Public Access Catalogues (OPAC), which have widely replaced the two former kinds, can be distinguished.

xIn literature [36], two types of cataloguing approaches are distinguished: formal cataloguing (usually simply referred to as cataloguing) and subject indexing. Formal cataloguing means to apply formal rules to describe books and other media using formal elements such as their author and title. These elements are inherently drawn from the media themselves. Therefore, formal cataloguing means to transform [30] data into a form compliant to rules. Older rule sets for doing so are

- the RAK (Regeln für die alphabetische Katalogisierung),
- the AACR (Anglo-American Cataloguing Rules) and
- the AACR2.

The new standard RDA (Resource Description and Access) for cataloguing introduced in 2010 has a broader scope and is aimed to be applied by museums and archives besides libraries as well. Furthermore, this rule set (https://access.rdatoolkit.org/) provides extensive guidelines to extract attributes of entities such as a particular edition of a book as well as to determine their relationships to other entities in order to support applications that rely on linked data.

On the other hand, subject indexing means to describe resources based on their contents and content-related criteria and without relying on bibliographic or other formal data. Thus, subject indexing means to interpret contents and therefore implicitly requires methods that can transform data into information. The two most common methods for doing so are keyword assignment and content classification. Keywords for a resource can be directly drawn from it or by relying on external contents such as reviews and annotations assigned by users. Categories make it possible to distinguish e.g. between person-, time- and location-related keywords. Content classification relies on a given, usually hierarchic classification scheme and aims to assign resources to categories and subcategories and therefore to ultimately group them based on their topical orientation. Both approaches can be applied together.

Aside from this rather formal and theoretic distinction of cataloguing approaches, the practical establishment of a library, which at its core means to turn textual data into information and ultimately knowledge, requires librarians to make considerable efforts and is definitely a time-consuming learning process in which the interaction with their users plays an important role, i.e. it is a process with a determined history. This implicitly means that two librarians ordering documents such as books may end up with completely different arrangements depending on their own knowledge gathered and the experienced process of knowledge acquisition.

It usually requires a deep study of the texts (if not even special knowledge on particular subjects) in order to find out important terms as well as to determine their context-dependent meanings which are subsequently to be used in the assignment of categories to previously unseen contents and in the determination of their relations. This process also involves an estimation of the semantic similarity and distance to other terms and texts locally available. Thus, only after a larger amount of knowledge is gathered, a first classification of documents may be carried out with the necessary maturity and a first, later expansible catalogue and archiving system may be established. The resulting catalogue is a small and compact abstraction of details in each book and in a condensed form even a representation of human intelligence that was used in connecting related books with each other and, in case of a card catalogue, in deciding on the card placements accordingly.

Technically speaking, this construction process follows—in contrast to Google's top-down approach—a bottom-up approach as the arrangement as well as the classification and sorting of books is carried out in a successive manner starting with an initially small set of books and is mainly determined by the specialised (local) and the common knowledge of the librarian. As the library is growing this way, its now existing classification scheme makes it easier to catalogue and order incoming books. Furthermore, it is – besides the librarian's own knowledge—the knowledge base for giving advices on where to find particular books or information of interest to library users. This approach is likely more expedient and successful than the mentioned top-down approach of Google and Co., especially when domain knowledge is needed to handle inquiries with particular terminology, literally speaking when it comes to finding the 'needle in the haystack' of information. As already pointed out in [76] (for the field of marketing), the usage of 'small data' (in the case at hand represented by the specialised knowledge of the librarian used to properly guide users to their requested information) is often more beneficial than to rely on improper big data analyses. Furthermore, based on this local knowledge, topically similar and related documents can be identified fast and are therefore typically assigned the same category in the library.

2.2 The Electronic Information Desk

The currently most important form of library catalogues is the so-called 'Online Public Access Catalog' (OPAC), an electronic bibliographic database, which has largely made the former physical card catalogues obsolete.

While OPACs make it possible for user to access and search for a library's resources using its respective online presence at any place and at any time, the integration and usage of Integrated Library Systems (ILS) [8] (which OPACs are a part of) has made the maintenance of these catalogues (management of metadata and information) along with the acquisition of media and management of loans more convenient for librarians, too. Especially, these systems make it possible to loan digital publications such as e-books, e-journals, e-papers (electronic newspapers and magazines) as well as digitised books and electronic course materials online.

Also, the cooperation between libraries has become easier by the introduction of data formats which foster the usage, exchange and interpretation of bibliographic information in records. For this purpose, the MARC (MAchine-Readable Cataloging) standard has been adopted widely. By this means, libraries cannot only offer their users local holdings, but can provide them records of associated libraries as well as additional services like inter-library book loans, too. This means that library users can access the preferred library's catalog locally (online and by physically going there) and are provided nationwide or even global information. The term 'hybrid library' [89] has been coined to indicate that a particular library offers classic and online services alike.

By the usage of ILS, spatial and temporal restrictions of classic libraries can be overcome as the provided electronic information desk is usually available around the clock. Thus, an adaptation to the users' communication behaviour is given. Furthermore, these systems support librarians by automatically analysing and forwarding user requests to the appropriate assistants. This assistance is even more extended by answering standard requests e.g. for opening times autonomously without the involvement of assistants. The integration of further electronic communication services such as chat and instant messenger services, microblogging sites, online social networks and Internet telephony has greatly facilitated the communication with library users. Online training courses can be easily provided and carried out by these means, too. Even so, it is always necessary to respect the protection of personal and private data, especially when a potentially large audience is addressed.

2.3 Searching the Web and OPACs

As mentioned in the introduction, web search engines can be helpful for finding relevant documents on short notice, especially when known items (e.g. the location of a shop) are searched for. However, when it comes to conducting comprehensive research on a topic of interest, users are—for the most part—not properly supported by them or even left for themselves. In such a case, usually referred to as subject search, users have to inspect the returned links to web documents by themselves, evaluate their relevance and possibly reformulate the initial or even subsequent queries in order to hopefully and finally satisfy their information need. This process is tedious and time-consuming alike, particularly when the user is unfamiliar with a subject and the proper terminology is yet unknown. Furthermore, most of the web documents are not catalogued or aimed for publication—in opposite to the literature—in libraries and as their trustworthiness cannot be taken for granted, it needs to be actively questioned at all times.

In these situations, libraries along with their services provided by both librarians and OPACs are definitely of more help. The reasons for this are obvious:

1. A library's literature has been intentionally selected and acquired.
2. A library provides literature in a well-ordered and structured form such that relevant contents in a field of interest can be found fast. A search process can be carried out in a more focussed way than it would be possible in a web search session.
3. Besides fields to formally classify a publication, OPACs provide dedicated fields for publications that are filled by applying methods of subject indexing. Faceted search is therefore a common feature in OPACs.
4. OPACs return a rich set of bibliographic information which simplifies continued searches of related materials. As an example, the author's name and the title of a publication are assigned to dedicated fields or elements with meaningful designations in the catalogue which makes it therefore easily possible for users to correctly interpret a publication's bibliographic information.
5. The information users are provided by libraries is usually trustworthy which includes both the literature found or suggested as well as other references to the subject of interest.

Subject searches are therefore more likely to succeed when relying on dependable library services. Furthermore, ILS are able to automatically extract citations from publications and link them to the literature referred to. The generated graph of related materials can then be the basis for content- or feature-based recommendation functions that users are accustomed to from e.g. web shops.

2.4 Decentralised Web Search with Library Services

Libraries have been early adopters of information technology. They applied information systems from the 1950s, a time during which the term 'information retrieval' (IR) [85] has been coined, too. At that time, professional searchers have been employed to act as 'search intermediaries' in order to translate users' requests into the respective system's language [135]. Nowadays, this function has been mostly replaced by search engines in various forms.

However, in order to realising a modern librarian-inspired decentralised web search engine as motivated in the introduction, the translation of users' information needs into their proper and promising technical representations as well as their matching with textual resources become again challenging core tasks of such a system. These tasks need to be carried out autonomously and automatically if necessary. Furthermore, in a decentralised setting, these representations must be routed and forwarded to peers that likely will be able to positively fulfil the mentioned information needs. Therefore, the next chapter gives an introduction to the general working principles of web search engines which covers the results of important research from the field of Peer-to-peer information retrieval (P2PIR), too.

The routing decision has to be made based on semantic considerations that also librarians would (unconsciously) take into account when guiding library users to relevant information and their sources. This is an especially crucial task as information in the web is largely unorganised as well as sparsely and often inconsistently annotated (if at all) by humans and machines alike and thus differs from information in catalogued library publications. In order to be able to do so, each peer of the proposed decentralised web search engine has to rely on a local knowledge base whose organisation closely matches the one of human (in this case the librarian's) lexical knowledge and therefore must be able to extract and index valuable information from textual sources and to put them into relation. This learning, ordering and cataloguing process can be implemented by applying specific algorithms and technical solutions known from the fields of natural language processing and text mining. The previously described cataloguing approaches can thus be applied in automatic form. However, the resulting catalogue or index is hardly comparable with rather monolithic, manually created OPACs as it is decentralised as well as automatically created and maintained. Therefore, relevant known as well as newly developed approaches in this regard will be presented in the Chaps. 4 and 7. Furthermore, in Chap. 5, previously developed and relevant approaches for local text analysis by the author of this book will introduced. They are particularly helpful for tasks such as automatic, high-quality key- and search word extraction.

At the same time, it is needed to account for implicit language-related dynamics of the web. Especially in online social networks and weblogs language change is recognisable. This does not only mean that topics gain or loose public interest during a specific time period and at specific locations but that the wording to describe them changes as well. Also, youth language and slang has to be dealt with accordingly. While librarians in particular are easily able to adapt to these changes due to their constant interaction with users of all ages and their growing knowledge of subject-related developments in the library's publications, these changes of meaning are up-to-now not properly accounted for by current semantic approaches for web search. Particularly, (usually) specialised (i.e. domain-related) ontologies or taxonomies cannot properly reflect language dynamics as they are normally manually created by humans, such as librarians, using a fixed terminology. Here, a new take on automatically handling these dynamics is needed:

1. A respective graph-based approach will be presented and evaluated in Chap. 6.
2. In Chap. 7, it is further used for the automatic clustering, classification, routing and forwarding of text documents (i.e. their cataloguing) and search queries alike.
3. Chapter 8 investigates its usefulness in search processes.

Especially, when it comes to conducting in-depth research in the web, the interaction with librarians is of great help due to their professional experience. In these cases, search becomes an information seeking process consisting of possibly numerous intermediate steps such as analysing presented information and reformulating queries. The proposed decentralised web search engine should be of similar usefulness in these situations. This means to interactively support the user in her or his current search task by giving instant feedback e.g. on the quality of a query or by suggesting (groups of) topically related query terms as well as by grouping similar and related web search results. In doing so, the system is able to learn from the interaction with its user and therefore to even make context-based predictions or to give recommendations on suitable next search steps. In this sense, an important step towards real 'information literacy' in information systems is taken.[1] The Chaps. 7 and 8 provide algorithmic solutions in this regard.

2.5 Summary

This chapter reviewed and classified the main services provided at libraries and the activities of librarians working there. Especially, their two main tasks of mediating between information and library users as well as of cataloguing the (incoming) library's resources have been described in detail. As these tasks are usually backed by electronic information and cataloguing systems, they have been elaborated on as well. Furthermore, it has been analysed which activities carried out by librarians can be technically realised in decentralised web search systems. Here, the reader is specifically pointed to the remaining chapters of this book as they are inspired by and closely correlated with these activities.

[1]The author is aware that real 'information literacy or competence' presented by humans is likely not to be reached by machines anytime soon.

Chapter 3
Contemporary Web Search

3.1 Measuring Relevance and Retrieval Quality

Classically, information retrieval (IR) systems (and web search engines make no difference) are designed to index and retrieve data from potentially large text document collections. These main functions are supported by methods to analyse queries as well as to preprocess incoming, search for and rank (partially) matching documents according to their estimated relevance. Information retrieval models such as the set-theoretic Boolean model, the algebraic vector space model and the probabilistic model [8] are commonly applied to automatically measure the documents' relevance to a given query. All of those methods neglect existing semantic relationships among the documents handled and terms involved as they perform their operations on token level only. However, newer models take into account term-dependencies such as Latent Semantic Indexing (LSI) [27] for doing so.

As these models are based on different approaches and their results returned may or may not satisfy the user's information needs, it is necessary to evaluate their retrieval quality. Common measures are

1. Recall R: the number of relevant documents returned out from the total number of relevant documents in the collection as presented in the following formula

$$R = \frac{|\{relevant\ documents\} \cap \{found\ documents\}|}{|\{relevant\ documents\}|}, \tag{3.1}$$

2. Precision P: the number of relevant documents among the number of all returned documents according to the next formula

$$P = \frac{|\{relevant\ documents\} \cap \{found\ documents\}|}{|\{found\ documents\}|}, \tag{3.2}$$

3. Traditional F-measure: the harmonic mean of recall and precision calculated by

M. Kubek, *Concepts and Methods for a Librarian of the Web*,
Studies in Big Data 62, https://doi.org/10.1007/978-3-030-23136-1_3

$$F = 2 \times \frac{P \times R}{P + R}. \tag{3.3}$$

These measures can be easily applied when so-called gold-standard datasets are available. They usually consist of a number of queries and documents that are for each query classified into the two categories 'relevant' or 'not relevant'. This assessment is based on human judgement. While those collections are undisputably useful, in a real-world scenario, the information needs and relevance judgements of humans depend on various context-related factors such as locations, time, mood and acquired knowledge. Thus, they are highly dynamic and not static as wrongly suggested by such collections. This is why, a number of approaches and models for interactive information retrieval [59] have been proposed that put the human (intelligence) into the centre of iterative information processing and retrieval.

Moreover, the calculation of recall is not easily possibly for large document collections (e.g. in the WWW) as it is usually unknown, how many document are available and how relevant they are to a particular query. A method, called pooling [78], can be applied to address these problems. Additionally, web search engines can evaluate the (user-dependent) relevance of results to a query by analysing the number of times they have been clicked. Especially when dealing with a web search system that possibly returns millions of search results, the individual user is usually interested in only the first few results pages. By inspecting them and by taking a look at content snippets usually provided along with the result links, the user is often able to instantly evaluate the quality of the (first k) results returned. This is why the measure Precision@k is of importance in these settings. It determines the precision of the top k results only.

In web information retrieval, the measure 'Normalized Discounted Cumulative Gain' (NDCG) [78] is now typically applied to access ranking quality given reliable relevance judgements. The approach is based on the notion of graded relevance as a measure of usefulness (gain) of a document in a ranked result list. Here, it is assumed that highly relevant documents are more useful than less relevant ones and that a relevant document is less useful for a user at a lower rank in the result lists as it is less probable that a user will recognise and inspect it. Therefore, the gain of a document at a lower rank is respectively reduced (discounted). In order to determine the overall gain of a result list at a particular rank k, the gains of the individual documents at their ranks i are summed up while starting from the top-ranked document. Formula 3.4 is used typically used for this calculation.

$$DCG_k = \sum_{i=1}^{k} \frac{2^{rel_i} - 1}{\log_2(1 + i)} \tag{3.4}$$

The relevance judgements rel_i for each document at rank i do not necessarily have to be provided in binary form (0 for not relevant; 1 for relevant). Instead, a grading scale can be used to distinguish between (highly) relevant and marginally relevant documents. In order to normalise the DCG at rank k, this value is divided by the DCG value at rank k of the ideal ranking (IDCG).

$$IDCG_k = \sum_{i=1}^{|REL|} \frac{2^{rel_i} - 1}{\log_2(1+i)} \tag{3.5}$$

Here, REL denotes the list of all available relevant documents (ordered by their relevance) up to position k. The number of all available relevant documents could be higher than the number of actually returned and relevant documents. The final value for the NDCG at rank k is obtained by:

$$NDCG_k = \frac{DCG_k}{IDCG_k} \tag{3.6}$$

The measure NDCG is useful when relevance judgements for all pairs of queries (to be tested) and documents are available. An ideal ordering, however, cannot be determined when these judgements are partially unavailable. A way to obtain them is—as mentioned above—to count the number of times users click on particular results. This implicit relevance feedback can provide valuable clues on which documents are relevant.

In contrast to traditional IR systems, centralised web search engines also have to deal with numerous challenges in order to return up-to-date results:

1. the amount of documents in the (surface) web to be (constantly) crawled and indexed is extremely large and growing and usually, the hidden or deep web cannot be crawled (see the next section on this point),
2. the documents' contents is dynamically changing and this change must be reflected in the search engine's index and search results,
3. their heterogeneity with respect to media types, different languages and character sets is high and
4. they contain a plentitude of language- and syntax errors (e.g. forgotten closing tags in HTML documents) which results in a varying data quality making an assessment of their trustworthiness hard or even impossible.

However, besides (natural) references, e.g. citations in scientific articles, between text documents found in the web, they contain explicitly set links in form of URLs (Uniform Resource Locators) to other and likely related web resources. These links are of benefit for both web users and the realisation of technical solutions alike. By means of links, users can instantly be made aware of related or recommended contents that can be simply reached by clicking on them. Additionally, linked contents are of relevance to the linking contents and these relationships can be technically exploited.

Even more interesting, links are helpful to analyse and technically use the web structure, too. In order to fill their indexes, web search engines e.g. follow them when crawling and scraping (extracting useful textual contents while filtering out advertisements and navigation menus from web pages) the web using breadth-first or depth-first traversal strategies while regarding the web sites' crawling policies.

Furthermore, algorithms to measure the relative importance of web pages such as PageRank [92] are able to do so by analying the global link structure found in the web. The obtained score is seen as one (among others) indicator for a content's relevance. The herein introduced technical solution for decentralised web search uses web links to automatically fill the list of neighbouring peers. This kind of natural structure building was specifically designed to address the bootstrap problem in P2P-systems.

In contrast to link- or solely text-based IR algorithms, content-based ones can also analyse and search for multimedia documents containing audio, video and picture data in various formats. Respective algorithms are subsumed under the term Multimedia Information Retrieval (MIR) [102]. Besides for their need to use particular technical approaches to handle the data elements' characteristics properly, textual metadata such as ID3 tags from music files can often be extracted from them as well and subsequently be fed to downstream text analysis algorithms which in turn can use this data for relevance computations as well.

3.2 The Architecture of Web Search Engines

Similarly to classic IR systems, web search engines have to index and return (links to) relevant web documents by matching them with incoming queries. The main question they try to answer for each query is: 'Where can a certain information be found?'. As this information is not implicitly available in the WWW, they have to perform a mapping from content (web documents with their words) to location (URLs of those documents). In this understanding, web search engines act as information brokers between users, contents and their locations. The technical infrastructure needed to carry out these three tasks consists of the following different main components [91]:

- the crawler,
- the local store,
- the indexer and
- the searcher.

Starting from an initially given set of URLs (seed set), crawlers (also called web robots or spiders) traverse the web graph by following links in web documents which it is made of. The visited web documents are then stored in a compressed form in the local store. Also, their links are extracted and stored. These links used to visit the next web documents linked to. While this approach is the most important one to fill the local store, due to the structure of the web [16], it is not possible to acquire all documents by this means. Therefore, it is necessary to properly fill the seed set in order to actually be able to reach a large portion of the web. Also, it must be noted that the crawlers can only reach contents that are publicly accessible (surface web). The so-called deep web, which consists of access-restricted contents or contents from databases only reachable using specific access mechanisms, cannot be crawled. Furthermore, webmasters can manually block certain contents from being crawled using specific instructions (Robots Exclusion Standard [56]) and document-specific meta information. However, there is no guarantee that web search engines obey these rules.

While the web search engines' general approach is to reach a high coverage of the web, vertical search engines (can also be a parts of web search engines) focus on documents of specific topics (often originating from particularly selected sources). On the one hand, focused crawlers are applied to gather these contents and create those topically focused collections. On the other hand, those collections can be filled by directly acquiring (structured) data offered by certain sources such as shopping portals and publishing houses. Due to the dynamic nature of the web, the local store must be kept up-to-date. This means that crawlers have to periodically visit especially frequently changing web pages to keep track of their changes.

The indexer's main task is to create and (exactly like the crawlers) continuously update the so-called inverted index which maps entries to the documents they appear in. For this purpose, it makes use of a parser module which generates those entries such as single words, their base forms as well as n-grams. Also, further characterising data for each entry is stored in the index like its frequency, font type and size and the position in the web document. Furthermore, the entries are stored in the particularly found form (capitalisation is maintained) and in lower case letters.

Additionally, document-specific metadata usually found in semi-structured HTML-files is stored in specific fields such as title, description, keywords, author, document length, the date of the last change and frequency of change. If particular fields such as the author have not been filled, they cannot be queried directly. In this case, however, the searcher could be used to carry out a full-text search for an author on token level. Furthermore, the indexer creates for each document an anchor file containing all document links with their absolute URLs (a URL resolver is used to transform relative URLs into absolute ones) and their anchor texts. The links in this file are stored in a link database which can be used by the algorithms given in the next section to determine the relative importance of web documents (see the next section) for document ranking purposes. The resulting index acts as a complex representation for the documents it has been created of which does not only store their textual contents in a technically manageable form but is enriched with useful meta information as well.

The searcher's task is to parse and interpret given queries and return links to relevant web documents found in the index. As these queries are expressions of the users' information needs, they often contain imprecise, short or ambiguous formulations. Therefore and in order to determine the most relevant documents properly, the searcher has to 'understand' the queries sent to it. For this purpose, contextual information is usually added to it such as information on the current user (her/his search history or location) or information on the behaviour of other users that have issued the same queries before, e.g. links they clicked on after they were presented their search results (click paths) and the amount of time spent for inspecting particular documents. Also, the current search session consisting of queries issued in a short amount of time can be used to disambiguate and/or enrich queries as they likely provide valuable contextual and topical clues. Furthermore, user profiles from online social networks that give an indication on the users' interests can be taken into account as well. However, query understanding is a nontrivial task as each final interpretation is only one of potentially many possibilities to (technically) make sense of a query's meaning. An inaccurate or even false interpretation would lead to a useless result set. In this regard, fuzzy logic has been used to interpret imprecise linguistic quantifiers in database queries [52]. Also, topic-related linguistic cues have been extracted and used to improve the effectiveness of interactive search [39].

After matching documents in the index have been determined, they need to be ranked according to their relevance. This crucial step is necessary as a large quantity of potentially relevant documents can and most likely will be found. Various ranking factors such as the number of matching query words, their distance as well as their positions in them in the documents are taken into account. Furthermore, the documents' relative importance in the web graph is usually factored in. Therefore, algorithms to determine this importance score are described in the next section.

Besides large-scale web search engines like Google (https://www.google.com/), Yahoo! (https://www.yahoo.com/) and Bing (https://www.bing.com/), there is a number of efficient and effective search solutions and software development frameworks for the creation of scalable search services available. As an example, the projects Apache Solr (http://lucene.apache.org/solr/) and Elasticsearch (https://www.elastic.co/de/) specifically address the needs of companies to e.g. realise web shops that are backed by full-text or faceted search capabilities. These two projects rely on the renowned Java-based text search library Apache Lucene (https://lucene.apache.org/). In combination with the projects Apache Nutch (web crawler, http://nutch.apache.org/) and Apache Tika (content extraction, http://tika.apache.org/) it is now easy to create an own basic but powerful web search engine within a matter of hours.

3.3 Ranking Search Results

3.3.1 PageRank

In order to present a ranked list of results, web search engines take into account the web's link structure to calculate a score indicating the relative importance of web documents. The famous graph centrality algorithm PageRank [92] measures the importance of a web page based on the importance of its incoming links. Although introduced in 1998, Google still applies PageRank as a factor to rank search results. The PageRank algorithm includes factors to cover the behaviour of a random web surfer. A user follows a random link on a web page with a probability of around 0.85 (damping factor d in formula 3.7) and visits another page not linked by this page with a probability of around 0.15. The PageRank of a page corresponds to the number of page visits of users. The more users visit a page the more important it is. For the PageRank calculation, the web pages are represented by nodes and the links between them are represented by directed edges, forming a directed graph. Besides other possibilities, the PageRank PR_i of a page i can be iteratively calculated by applying the following formula:

$$PR_i = (1 - d) + d \sum_{j \in J} \frac{PR_j}{|N_j|}. \quad (3.7)$$

The Set J represents the pages linking to page i and $|N_j|$ is the outdegree of page j. In the last years, many solutions for distributed PageRank computations [47, 111, 141] have been published in order to address the shortcoming that originally the whole web graph needs to be considered. In [114], extended methods for distributed PageRank calculation including network parameters based on random walks are discussed and empirically evaluated. Another link-topological algorithm that can be applied in distributed or decentralised systems is called NodeRanking [113]. This algorithm is particularly interesting for application in P2P-systems as it allows to rank nodes using local information only.

However, besides PageRank, according to [26, 37], Google applies many other factors or signals to rank search results.

Such signals are provided by its core algorithms [43] 'Panda' (assesses content quality), 'Penguin' (assesses link quality) and 'Hummingbird' (interprets queries correctly). In 2015 [21], the artifical intelligence system 'RankBrain' has been introduced as a part of 'Hummingbird'. It delivers now one of the most important ranking signals. The system has been conceived to better interpret previously unseen queries such as precise questions. This way, relevant documents can be found that do not necessarily contain words that were used in the queries. By means of its learning and generalising capabilities, 'RankBrain' can even 'guess' the correct meaning of colloquially formulated or ambiguous queries by recognising similarities to previously understood, related inquiries. Summarising, ranking search results means today to apply machine learning algorithms on a large number of features (signals) most of which stem from query logs, i.e. usage data.

3.3.2 Hyperlink-Induced Topic Search (HITS)

The HITS algorithm [55] bears similarity to the PageRank algorithm. It was initially designed to analyse the link structure of web documents returned as results from web queries and to determine their relative importance. The algorithm returns two lists of nodes (representing those web documents): authorities and hubs. Authorities are nodes that are linked to by many other nodes whereas hubs are nodes that link to many other nodes. Nodes are assigned both a score for their authority and their hub value.

The authority value of node x is iteratively determined by the sum of all hub scores of all nodes v linking to it:

$$a(x) = \sum_{v \to x} (h(v)). \tag{3.8}$$

Its hub score is determined by summing up all authority values of all nodes that it links to:

$$h(x) = \sum_{x \to w} (a(w)). \tag{3.9}$$

For undirected graphs, the authority and the hub score of a node would be the same, which is naturally not the case for web (sub)graphs. However, this is not the only application field for this algorithm. As shown in Sect. 4.4.3, the HITS algorithm is of use in graph-based text mining solutions, too.

3.3.3 Other Ranking Algorithms

Further context-sensitive algorithms such as the Page Reputation algorithm [101] and Topic-sensitive PageRank [41] both extend the PageRank algorithm and take into account the user's particular interest for given terms. Thus, web pages receive a higher score if they actually contain these terms. However, there is the notable risk of receiving topically biased search results when applying these approaches.

3.4 Interactive Web Search

Despite recent advances in query formulation and interpretation techniques, the most common way for users to search for information in the WWW's vast amount of data is to express their information needs by entering search words into the query input fields of web search engines. The search words are chosen such that they are likely to appear in documents that are expected to fulfil those information needs. Here, the challenge is to describe desired contents with only a few keywords expected to be found in the contents and thus to reduce the effort needed to input multiple query terms. In average, only two words [2] are used to address the wanted content. As these short and imprecise queries often contain ambiguous terms, this approach usually results in a large number of documents presented to the users, who mostly review only the first 10 to 30 results [74]. Ranking algorithms are applied to ensure that only links to the most relevant search results appear among those first 30 result entries. The heatmap presented in Fig. 3.1 indicates the area of result lists that is typically skimmed through first by users.

In general, the search for nontrivial information in the web is an iterative process. In the first step, the user formulates a query depending on her or his information needs and sends its search words to the search engine of choice which—as already pointed out—enriches it with contextual information and returns a list of matching

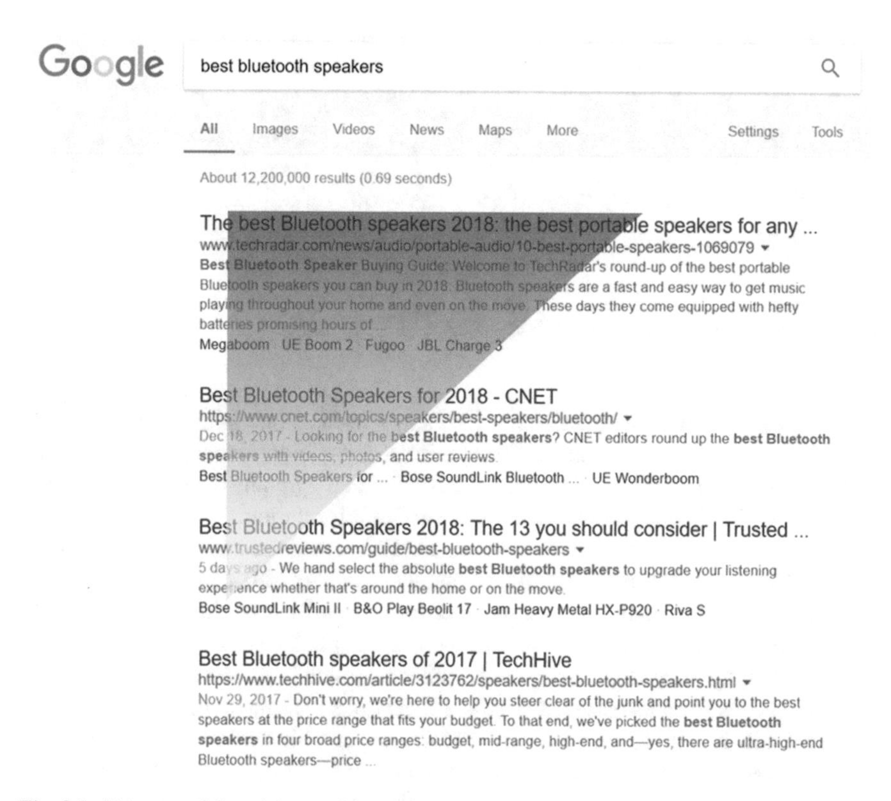

Fig. 3.1 Heatmap of Google's search results

documents. Content snippets of some results are quickly scanned and an even lesser number of them are inspected afterwards in order to specify the wanted information more precisely and to learn about more specific keywords from the target topic. Then, the newly found terms are possibly used for another request and the described process continues until the desired documents are found or the search was finally unsuccessful. Depending on the particular experience of the user, this process may take more or less time. Especially for non-experts and novices in a field, it may become time-consuming, error-prone and tedious, when e.g.

- the amount of returned search results is (despite the application of ranking algorithms) too large,
- the search results cover too many topics or subtopics,
- the user lacks sufficient knowledge on the topic of interest and
- when fully new or emerging topics are of interest for which there might not yet exist a proper terminology.

An automatic recommendation of search words and queries that regards the context of the search subject would therefore be of great help to the user. To this end, most search engines upgraded and extended their services by an introduction of user

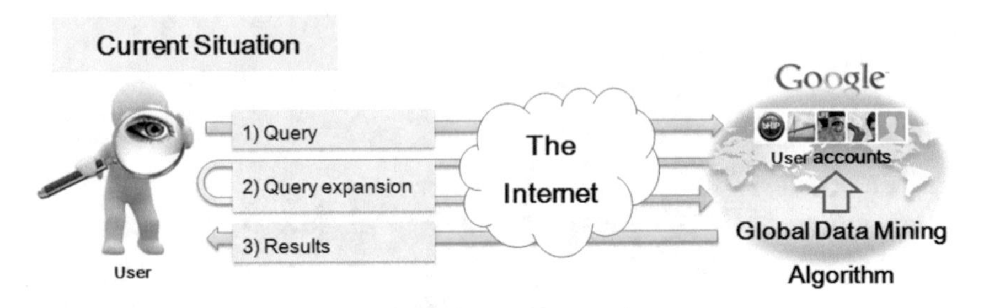

Fig. 3.2 The process of searching the web

accounts and statistical evaluations of frequently used search phrases and n-grams. In such a manner, search engines offer their users additional query terms to refine their search based on the keywords previously entered by many users along with the initial query [129], geographic location or profile information.

By using these services, however, the users often provide sensitive personal details related to their private lives. By this means, the search engine can collect and store a bigger set of information on the respective user and learns details about her or his special behaviour and interests. Google itself applies a statistical approach and offers high frequency n-grams from document contents as well as frequently co-occurring terms from its users' queries as suggestions to refine queries. This method of course fails if completely new, very seldom or special-interest content is searched for. However, information obtained about the user is often used for other purposes than just for the improvement of web search, e.g. to place targeted commercial advertisement based on the area of his or her search activities. The common process of searching the web (only one iteration) is depicted in Fig. 3.2.

The acceptance of these suggestions can be regarded as a positive implicit relevance feedback as they help users in reformulating their queries to better match relevant documents. Although the concept of explicit relevance feedback is known since 1971 [107], it is not implemented in the current web search engines' user interfaces and search algorithms. However, it would be a tremendous advancement if users could filter out e.g. groups of topically irrelevant results by simply 'removing' one representing document from the result list or by directly using query terms from documents marked relevant for further search iterations.

3.5 Semantic Approaches to Web Search

Mostly, web search engines carry out their search tasks by matching queries and documents on a simple token-level basis. Despite their usefulness to e.g. disambiguate queries, semantic aspects are usually not considered in doing so. As a first approach to take them into account as well, Google itself introduced in 2012 its so-called 'Knowledge Graph' [38], an entity-centric knowledge base which has been applied as a search-augmenting technology. This knowledge base is used to enrich search results with relevant links real-world entities such as information on persons and their relations. The fact (search) engine 'Wolfram|Alpha' https://www.wolframalpha.com/ considers itself a 'computational knowledge engine' and is able to semantically interpret queries by performing linguistic analyses, draw logic inferences from facts and entities found and provide useful information on their relationships using its large factual database. The roots of the wish to make web contents explicitly machine-readable and semantically findable are, however, much older.

In 2001, Tim Berners-Lee coined the term 'Semantic Web' [10], which was frequently discussed among experts in the field of knowledge engineering. To reach this goal, entities or resources (referenced by their URI (Uniform Resource Identifier)) and their relationships found in web documents are modeled using the Resource Description Framework (RDF) or more expressive ontologies generated using the Web Ontology Language (OWL) [124]. With RDF, it is possible to make statements in form of triples which express a particular relationship (predicate) between a subject and an object. These triples are typically stored in an XML-document. A set of linked triples is called an RDF-graph which can be searched/traversed using specific query languages such as SPARQL [125]. This approach makes it possible to automatically infer further information on resources from these graphs and to annotate web pages with these relationships. However, in order to fill these graphs, facts and their relationships must be automatically extracted from natural language text found in the WWW by relying on text mining methods such as named-entity recognition and information extraction. A manual acquisition is infeasible due to the large amount of textual data to be analysed. This problem is even more pressing when ontologies (typically for a specific domain) shall be used.

First, ontologies (despite the fact that there are approaches for learning ontologies from web documents such as YAGO [103]) are manually created by domain experts in often tedious processes. Incompatibilities between ontologies are therefore to be expected, even within specific domains.

Second, the resulting ontologies model represents dependencies of and relationships between concepts which need to be automatically assigned proper real-world facts and entities in order to actually make them useful in search applications.

Due to these problems, the 'Semantic Web' is up to now a matter of academic research. In industry, these approaches do not play a significant role yet. However, there is a growing interest [105] from companies to integrate those technologies to improve their knowledge management and e-commerce solutions. Returning back to search applications, the semantic search engine NAGA [53] as an example relies on YAGO [103] as a knowledge base (modeled as a fact graph) and offers an expressive query language that allows to formulate precise discovery and relatedness queries. The returned results are ranked/weighted according to an overall certainty value derived from the certainty values assigned to the involved facts that have been extracted by potentially unreliable information extraction methods.

Another important term in this context is 'Web 2.0' [90]. Here, the focus is set on the users. The aim is to improve socially driven cooperation and the easy creation and exchange of information in the WWW. Services like online social networks,

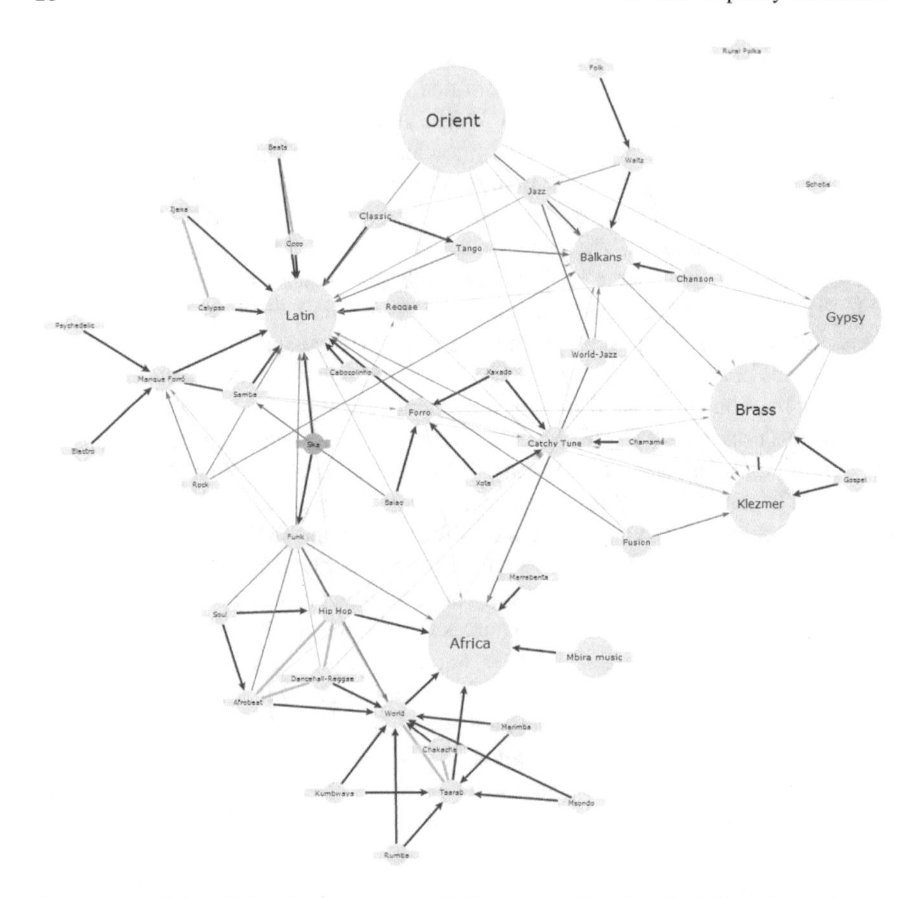

Fig. 3.3 Graph-based visualisation of semantically connected music-oriented tags (genres)

weblog systems and the popular online library Wikipedia are some examples make the benefits of collective intelligence provided by their users visible. It is also useful for the collaborative classification of items such as songs, pictures and bookmarks using freely chosen tags. As many people take part in tagging items, frequently chosen tags clearly indicate proper semantic annotations. They are also potentially useful candidates for identifiers or resource descriptors in the 'Semantic Web'-approach.

Thus, tags from well-known items can be suggested as classifications for new (possibly still untagged) and semantically or content-wise similar items. Popular services that apply this social tagging approach are Delicious (https://del.icio.us/), Flickr (https://www.flickr.com/) and Last.fm (https://www.last.fm/).

Usually, items often have been tagged with more than one annotation. Co-occurring annotations can be assigned a weight or significance value with respect to their individual frequency (per item) in the whole system. This significance value can be interpreted as a measure for the surprise/coincidence that these annotations occurred together. Semantically related annotations will receive a high significance relaxing Send Query

(Related classes significantly co-occurred with the query term. The greater the significance value is, the more related the classes are to the query term. Indirectly related classes significantly co-occurred with the related term(s) and therefore are semantically connected to the query term as well.)

Related to relaxing:	Significance:	Indirectly related to relaxing via first column:
soft	100.023446343815	romantic calm africa guitar sensual contrabass mbira mbira music cello
calm	82.5879867696858	melancholic soft romantic change of mood tamburitza yearning classic waltz sensual
tango	35.8092628849599	coffeehouse-music background music sensual cello (three celli) classical classic balkans melancholic yearning
africa	26.2566424986444	mbira music mbira marimba soft summer spiritual world marrabenta guitar
lounge	25.551010189135	sensual beats cavaquinho electronic tango samples melancholic yearning coffeehouse-music
background music	23.3677370611031	tango coffeehouse-music sensual balkans melancholic cello (three celli) passionate yearning classic
coffeehouse-music	18.9081390344782	tango background music cello (three celli) sensual classic classical balkans lounge piano
sensual	17.5475158055533	kanoun tango strings (arabic) yearning background music coffeehouse-music melancholic bellydance romantic nay
mbira	17.3965557996407	mbira music africa marimba spiritual ruminant soft summer
mbira music	16.6874197110565	mbira africa marimba spiritual ruminant hosho soft ngoma summer

Fig. 3.4 Recommendation of related tags

value. Furthermore, as seen in Fig. 3.3, the graph-based visualisation of those relations can give interesting insights into the context of particular annotations of interest. The concept of co-occurrences in the field of automatic text analysis will be elaborated on in the next chapter in detail.

In order to validate this approach, the author has developed a recommender system that can suggest directly and indirectly related music-oriented tags to given queries. As shown in Fig. 3.4, the semantically closest tags to the query 'relaxing' are 'soft' and 'calm'. Also, indirectly related and possibly useful tags such as 'romantic' and 'melancholic' have been found. In order to present this kind of suggestions, 1.2 million statistically significant relations between 2800 music-oriented tags (using the log-likelihood ratio as a significance measure) of the 1500 most popular songs of 2009 in Germany have been extracted from a Last.fm dataset. Tags have only been taken into account, when they were assigned at least five times to different items in the dataset. This example shows that statistical co-occurrence analysis is also helpful for the realisation of recommender systems, even when multimedia data such as music, pictures and videos need with no explicit textual descriptions need to be suggested. The only prerequisite to successfully apply this approach is that this data is properly annotated. This kind of community-based classification can also be used to complement content-based approaches which often return results that do not represent the perception of humans (semantic gap). Content-based classifiers should therefore be able to learn from human annotations provided and thus be able to adapt their parameters accordingly.

3.6 Critique of Today's Web Search Engines

When it comes to simple query processing, connectivity analyses and statistical computations that current web search engines can carry out using their huge data bases, they greatly outperform any librarian, even when taking into account the rather limited amount of information in the books in a single library.

However, the establishment of bibliographies and topical signposts are services of a librarian no search engine can offer so far, since every document is considered as an own instance and only relations between documents using links in them are considered. Finally, the author argues that search engines cannot compete with the human classification process yet, because they do not support the already described maybe slow, but supervised learning and evaluation process. They always apply the same standard token-based techniques such as keyword processing to index all crawled documents, whereby semantic dependencies within or between documents processed do not play a big role. This also means that topically related or close resources (e.g. as helpful clusters) are not presented as such (aside from specialised services like Google Scholar (https://scholar.google.com/)). Advanced research tasks in the humanities such as tracking topics (to their roots) are hard to carry out by using them. The outcome is a linear and ranked search result list which does not reveal any semantic inter-document dependencies. All those problems may be solved however by hard working and experienced librarians.

It has been pointed out in several publications like [137] and [71] that query expansion may reduce the amount of search results significantly by a better description of the search subject. Therefore, most search engines already provide such kind of suggestions to their users based on the keywords entered by many users along with the initial query [129]. Furthermore, besides HTTP cookies, a user account at one of the big search engine providers is the basis for personalised query suggestions and tailored result lists. In doing so, the search engine learns about the special search behaviour and interests of its users. Of course, this approach has another aspect: the user discloses personal details related to private life that may be analysed, too. As individual persons will regard different results as relevant, this approach might seem generally sensible. However, these targeted results can also lead to so-called 'filter bubbles' [94], in which the users are prominently shown results that support their particular opinions or viewpoints while at the same time opposing contents are ranked low. The term 'echo chamber' [49] refers to the same phenomenon, but is more negatively [75] connoted. Filter bubbles pose a risk to researchers in particular that are usually interested in unbiased information. In that regard, a topical grouping of results would lower the impact of this effect as users are presented possible topical directions to follow at first glance.

Information literacy [3], i.e. the ability to recognise when information is needed and to have the ability to (proactively) locate, evaluate, and effectively access needed information, is not demonstrated by Google and Co. as it is not implemented. Also, it is hard for their users to gain it by using them, as the presented search results are merely pointers to documents with potentially relevant information. Usually, these linked documents must be manually inspected and their relevance to the current information need evaluated. In this regard, the presented content snippets are helpful for a superficial evaluation at first glance. However, as the document links are not topically grouped or classified, this inspection can turn into a elaborate task.

Summarising, from the user's point of view, web search engines have a number of problems that are not properly solved yet:

1. although the ranking of search results is generally satisfying, the amount of results returned is too large (user is overloaded),
2. search results are not topically grouped,
3. personalised search results require a user account (this may present privacy issues as users disclose personal information, which can be used to identify them),
4. 'filter bubbles' occur making it hard for users to discover unseen, yet relevant contents,
5. even so, the individual user's knowledge is not sufficiently considered when compiling search results.

From the technical point of view, centralised web search engines generate a findable connection between contents and their locations by storing a copy of the entire web in their indexes. As mentioned in the introduction, this brute-force method of making the web searchable is characterised by a significant overhead for maintaining and updating them. Furthermore, the used technical components like servers and databases are potential targets for cyber-attacks and pose a threat to the system's safety and security as well as to data protection.

Therefore, a new concept for web search is needed which shall be motivated in the next section by referring to related preliminary works.

3.7 Decentralised Approaches to Web Search

It is urgently necessary to properly address these problems of both centralised web search engines and semantic approaches. As mentioned in the introduction, several requirements for the search in tomorrow's Internet can be derived from these shortcomings. This book therefore introduces a new concept along with its technical solutions and infrastructures for future, decentralised web search while considering these requirements. Hereby, the focus is set on the emergence of brain-like and physics-inspired structures based on learning and adaptation processes that require no central control or knowledge but take into account feedback from various multiple sources. To put this line of thinking in other words: if 'the network is the computer' (John Gage, SUN, 1984), which shall 'learn from examples' [140] and the behaviour of the user to restructure itself, this goal can be achieved.

3.7.1 Historical Aspects of P2P-Based Search

First tries to avoid centralised instances and to build a fully decentralised Internet service system dated back to 1997 with the introduction of the Web Operating System ($WOS^{(TM)}$) [57]. It realised a remote service execution with a search service using local warehouses containing a set of privately known neighbours of every node (and therefore forming a connected service network) and cooperative services of any other network member. Later, around 1999, this idea has been made perfect in the peer-to-peer (P2P) paradigm, mostly used within file-sharing systems like *Gnutella* or *FreeNet* [97] or content-delivery services [113]. Peers are, in contrast to the client-server model in which there exist only clients requesting services usually offered by web servers, computer programs that can act as both clients and servers (usually at the same time). This way, they can offer own and make use of remote services provides by other peers. Differing from centralised approaches, a peer only has local as well as some limited information on its neighbours. Traditionally, two kind of P2P-systems are distinguished: structured and unstructured P2P-systems.

One of the most common use cases for these systems is to find contents or information by issuing queries. The thereby generated number of messages in unstructured P2P-systems like the ones mentioned before are bounded by two mechanisms: a time-to-live (TTL) or a hop-counter determining the maximum number of forwards for any message (usually set to 7) and the rule that every message (identified by a unique message-ID) is only forwarded once by every node (no matter how often it has been received from different neighbours). Consequently, in a complete graph, the number of messages cannot exceed N^2, if N denotes the number of nodes. In technical systems like *Gnutella*, every node usually keeps only a limited number of random connections open (no predefined network structure is imposed), e.g. 4. In this case, the maximal number of messages is limited by $4 \cdot N$ even, also when flooding-like (restricted by the mentioned number of connections) forwards are carried out. While the number of messages linearly depends on the size of the network, the average distance between a desired information and the requesting node is usually lower than $\mathbf{O}(ln(N))$, as it is known from the works on small-world graphs initiated by Stanley Milgram in 1969 in [83]. Furthermore, so-called super peers (computers equipped with many computational resources) can be used to control the network structure and to optimise routing as well. Further advantage of those unstructured networks is that their maintenance is computationally cheap.

Structured P2P-systems like CHORD, PASTRY or CAN (just to mention a few; for an overview see [20]) made the search in decentralised systems more efficient by limiting the average expenditure of a search to $\mathbf{O}(ln(N))$ but generated expensive overhead to maintain those structures in case of leaving and joining peers. They also do not support multi-keyword searches per se as for each single query term the search has to be carried out separately. Therefore, their practical use in decentralised information retrieval systems is limited. However, for the first time, the importance of specialised structures and their self-organisation/emergence have been underlined by those systems.

3.7.2 P2P Information Retrieval

In order to use P2P-systems for the purpose of information retrieval, one has to keep in mind that—in contrast to the use case of content delivery—replica of (relevant) documents often do not exist. Thus, it is needed to find the few peers that actually can provide them. Therefore, efficient routing mechanisms must be applied to forward a query to exactly those matching peers and to keep network traffic at a low level. But in order to be able to do so, a suitable network structure must be set up and adapted in a self-organising manner as well. At the same time, such a network must be easily maintainable.

Some of the most important results in the field of P2P information retrieval (P2PIR) have been obtained in the projects SemPIR I and II, financed by the German Research Foundation (DFG) from 2004 to 2009 [130]. Their goal was to make search for information easier in unstructured P2P-networks. In order to reach this goal, a self-organising semantic overlay network using content-depending structure building processes and intelligent query routing mechanisms has been built. The author actively participated in these projects. The basic idea of the approach applied therein is that the distribution of knowledge in society has a major influence on the success of information search. A person looking for information will first selectively ask another person that might be able to fulfil her or his information need. In 1967, Milgram [83] could show that the paths of acquaintances connecting any two persons in a social network have an average length of six. These so-called small-world networks are characterised by a high clustering coefficient and a low average path length. Thus, the mentioned structure building processes conceived modify peer neighbourhood relations such that peers with similar contents will become (with a high probability) direct neighbours. Furthermore, a certain amount of long-distance links (intergroup connections) between peers with unrelated contents is generated. These two approaches are implemented in order to keep the number of hops needed (short paths) to route queries to matching peers and clusters thereof low. This method is further able to reduce the network load.

In order to create those neighbourhood relations, a so-called 'gossiping' method has been invented. To do so, each peer builds up its own compact semantic profile (following the vector space model) containing the k most important terms from its documents which is periodically propagated in the network in form of a special structure-building request, the gossiping-message. Receiving peers compare their own profiles with the propagated one and

1. put the requesting peer's ID in the own neighbourhood list and
2. send the own profiles to the requesting peer

if (depending on a threshold) the profiles are similar to each other. Also, the requesting peer can decide based on the received profiles which peers to add to its neighbourhood list. Incoming user queries (in the form of vectors as well) are matched with the local profile (matching local documents will be instantly returned) and the profiles of neighbouring peers and routed to the best matching ones afterwards. This mechanism differs from the mentioned approach in real social networks: in the technical implementation, the partaking peers will actively route queries from remote peers. In real social networks, people will likely just give the requesting person some pointers where to find other persons that have the required knowledge instead of forwarding the requests themselves.

In doing so, a semi-structured overlay P2P-network is built which comprises of clusters of semantically similar peers. This way, queries can be quickly routed to potentially matching target peers that will likely be able return useful contents. Additionally, each peer maintains a cache of peers (egoistic links) that have returned useful answers before or have been successfully forwarded queries to matching peers. Furthermore, the network's structure is not fixed as it is subject to dynamic changes based on semantic and social aspects.

3.7.3 Further Approaches and Solutions

Further approaches to P2P-based search engines are available, too. *YaCy* (https://yacy.net/de/index.html) and *FAROO* (http://www.faroo.com/) are the most famous examples in this regard. However, although they aim at crawling and indexing the web in a distributed manner, their respective client-sided programs are installed and run on the users' computers. They are not integrated in web servers or web services and thus do not make inherent use of the web topology or semantic technologies for structure-building purposes. Especially, they do not take into account semantic relationships between documents. In contrast to these approaches, *Edutella* (https://sourceforge.net/projects/edutella/) is a peer-to-peer network for the search and exchange of content descriptions (especially educational metadata) in RDF. Relying on the (now discontinued) open source peer-to-peer protocol JXTA (http://www.oracle.com/technetwork/java/index-jsp-136561.html), Edutella offers a distributed query and sharing service for this purpose. It is to be noted that contents are not shared but their metadata only. In such a manner, P2P systems could become an excellent foundation of a new generation of Internet search systems. To this approach, other works contribute as well that e.g. find clusters of nodes [121] and optimise their neighbourhood depending on content aspects [22, 109]. Meta-searches [80], the use of previous searches and results (as well as caching) [7] and approaches to collaborative filtering and search [95] complete the set of already realised approaches to improve the search in the Internet. Last but not least, some efforts have been put into investigations in the area of ontologies [127], content annotations [116], web search with text clustering [132], probabilistic methods [46], deep semantic analysis [44] as well as brain-like processing [42].

3.7.4 Requirements for the Librarian of the Web

All of these contributions have influenced the work of the author and have therefore been mentioned here. In the following chapters, a basic, new concept for decentralised web search, subsumed under name 'Librarian of the Web', as well as its technical implementations shall be derived and explained. Based on and in continuation to the foregoing considerations and identified shortcoming of current web search engines, the author argues that such a new kind of librarian-inspired information system for the WWW should not only offer the aforementioned library-specific functions, but replace the outdated, more or less centralised *crawling-copying-indexing-searching* procedure with a scalable, energy-efficient and decentralised *learn-classify-divide-link & guide* method, that

1. contains a learning document grouping process based on a successive category determination and refinement (including mechanisms to match and join several categorisations/clusters of words (terms) and documents) using a dynamically growing or changing document collection,
2. is based on a fully decentralised, document management process that largely avoids the copying of documents and therefore conserves bandwidth,
3. allows for search inquiries that are classified/interpreted and forwarded by the same decision process that carries out the grouping of the respective target documents to be found,
4. is (at least at its core) language-independent,
5. ensures that the returned results are 100 percent recent,
6. supports a user-access-based ranking to avoid a network flooding with messages,
7. returns personalised results based on a user's locally kept search history yet does not implicitly or explicitly propagate intimate or personal user details to any centralised authority and therefore respects data privacy and contributes to information security and
8. returns results without any commercial or other third-party influences or censorship.

Differing from the approaches cited above, the author intends to build and maintain a P2P-network (the actual realisation of the 'Librarian of the Web') whose structures are directly formed by considering content- and context-depending aspects and by exploiting the web's explicit topology (links in web documents). This way, suitable paths between queries and matching documents can be found for any search processes.

3.8 Summary

This chapter provided an overview of existing approaches and solutions for contemporary web search while distinguishing them from classic information retrieval and approaches for semantic search. The general architecture and working principles of current web search engines have been explained. Furthermore, their drawbacks and problems from the technical and from the user's point of view have been identified and elaborated on. The herein presented concept and technical solutions for future decentralised web search, subsumed under the name 'Librarian of the Web', address these problems on multiple levels. As their central task is to deal with a large amount of natural language texts in an unsupervised manner, in the next chapter, basic approaches for natural language processing as well as methods for advanced text mining are discussed. Here, the focus is set on techniques to determine similarities between words and texts as well as to group them based on semantic aspects.

Chapter 4
Natural Language Processing and Text Mining

4.1 Brain-Inspired Text Understanding

In order to technically realise the 'Librarian of the Web', it is useful to consider the continuous process of knowledge acquisition of the human brain by gathering, classifying and combining information, which seems to be a strict sequential learning process [42]. Starting with the study of a first document, the brain gets an impression of its keywords, their frequencies but mostly of regularities such as frequent and joint appearances of words in particular units of texts (not necessarily sentences). This way, commonly co-occurring terms/phrases are learned and a 'feeling' for the general language use is gained [118] which is in turn needed to understand unseen content and a prerequisite for language production as well. Any further text is unconsciously and semantically compared with recalled contents of previously read texts, resulting in a procedure that determines.

1. how important particular terms/words/phrases are for the general understanding of the text's meaning,
2. which words are of no importance for the contents (stop words),
3. if and to what extent particular terms/words/phrases relate to each other semantically and
4. ultimately to what extent there is a similarity between two texts (this part of understanding the meaning/semantics of a text is hard to transfer to a machine)?

As the main task of the 'Librarian of the Web' is to manage and topically group text document collections, it is important to programmatically identify characteristic terms (e.g. keywords, concepts), words and word forms from those documents and to determine their semantic relatedness as well. The problem arising in this context is that word forms are basic elements in texts and, therefore, carry no explicit attributes to characterise their semantic orientation which could be used to compare their similarity. This is why, in the following sections, techniques to solve these problems shall be presented.

M. Kubek, *Concepts and Methods for a Librarian of the Web*,
Studies in Big Data 62, https://doi.org/10.1007/978-3-030-23136-1_4

4.2 Basic Tasks in Automatic Text Analysis

As natural language text is unstructured data (from the machine's point of view), it needs to be cleaned and transformed into information or another representation of it that can be uniformly handled by algorithms. That is why linguistic preprocessing is usually applied in text analysis pipelines [126] on the documents to be analysed whereby

- running text from text files in different file formats (media types) is extracted while also removing structural elements from e.g. HTML-files such as tags, menus and advertisements,
- the used language is identified,
- sentence borders are detected,
- individual tokens (word forms) are determined by considering the identified language, punctuation marks, abbreviations and contractions,
- stop words (short function words from closed word classes that carry no meaning) are removed (this optional step depends on the intended application),
- word forms found are tagged with their particular part of speech and
- the base or root form of them is determined by stemming or the more complex lemmatisation (typically, nouns, proper nouns, names and phrases are extracted for further considerations; however, adjectives and verbs in the field of sentiment analysis are useful, too).

Usually, the finally extracted words actually determine the meaning of texts as they represent real-world entities, their properties as well as actions that concern them. In this understanding, a word is an equivalence class of related word forms (inflected form of a given root word) usually found in texts. To simplify the following elaborations, the words 'term' and 'word' are used synonymously and mostly concern nouns, proper nouns and names. Phrases containing selected parts of speech will be specially mentioned when needed as well. It is assumed that the terms/words referred to are available in their base form. This implies that specific word forms are not considered as they occur in the text as linguistic preprocessing has been applied on them. There is a large number of actively supported software applications and libraries available that can reliably carry out these basic tasks in natural language processing (NLP). Some of them are:

- Apache OpenNLP (https://opennlp.apache.org/),
- GATE (https://gate.ac.uk/),
- Stanford CoreNLP toolkit (http://stanfordnlp.github.io/CoreNLP/) and
- LingPipe (http://alias-i.com/lingpipe/).

It is noteworthy that many NLP solutions are Java-based as they originate from academic institutions which often teach Java as a first programming language.

After linguistic preprocessing has been carried out, usually, the steps of term frequency analysis [46] and term weighting (e.g. using TF-IDF [8]) are applied. These steps still operate on token-level and are approaches to answer the first two questions given in a simplified way, whereby the overall occurrence frequency of individual words in the text is considered, but not word sequences or frequently occurring word patterns or relations.

The determination of significant collocations [100] and co-occurrences [46] using statistical analysis as well as the application of methods such as latent semantic indexing, (LSI, see [27]) which uses linear algebra techniques to assign related terms to one and the same concept in a lower dimensional concept space, are approaches to address the third question. The most prominent solution to answer the last question involves the usage of the vector space model known from information retrieval.

These and further techniques are of great importance to determine the relative importance of terms/words, to quantify the strength of their relatedness and to ultimately assess the semantic similarity of texts. They will be explained step-by-step in the next subsections. Moreover, these techniques are particularly worth mentioning as they are not only the core building blocks to realise a variety of text mining applications such as document clustering solutions but also enable the creation of sophisticated search applications. Hereby, state-of-the-art graph-based techniques shall be highlighted as their good performance in numerous applications in the fields of natural language processing and text mining have been confirmed.

4.3 Identifying Characteristic Terms/Words in Texts

The selection of characteristic and discriminating terms in texts through weights, often referred to as keyword extraction or terminology extraction, plays an important role in text mining and information retrieval. The goal is to identify terms that are good separators that make it possible to topically distinguish documents in a corpus. In information retrieval and in many text mining applications, text documents are often represented by term vectors containing their keywords and scores while following the bag-of-words approach (the relationship between the terms is not considered). Text classification techniques as an example rely on properly selected features (in this case the terms) and their weights in order train the classifier in such a way that it can make correct classification decisions on unseen contents which means to assign them to pre-defined categories.

As a first useful measure, variants of the popular TF-IDF statistic [8] can be used to assign terms a weight in a document depending on how often they occurs in it and in the whole document corpus. A term will be assigned a high weight, when it often occurs in one document, but less often in other documents in the corpus. However, this measure cannot be used when there is no corpus available and just a single document needs to be analysed. Furthermore, this statistic does not take into account semantic relations between the terms in the text.

Keyword density [96] is another measure which determines the percentage of times a term occurs in a specific document compared to the total number of terms in this document. This simple approach works for single documents but assigns term weights without considering term relevancy. Moreover, terms will receive a high weight just because they appear very often in a document which is a major drawback as stop words, general terms and a maybe high number of maliciously placed terms in texts will be illegitimately granted high importance.

The so-called difference analysis [46, 136] from the field of statistical text analysis is able to find discriminating terms. Terms in a text are determined and assigned a weight according to the deviation of their relative frequency in single (possibly technical) texts from their relative frequency in general usage (a large topically well-balanced reference text corpus such as a newspaper corpus which reflects general language use is needed for this purpose). The larger the deviation is, the more likely it is that a (technical) term or keyword of a single text has been found. If such a reference corpus is not available, this method cannot be used.

Under the assumption that a weight for terms even in single texts needs to be determined, it is therefore sensible to consider the semantic relations between terms and to determine their importance afterwards. Approaches following this idea would not require external resources such as preferably large text corpora as a reference. Two state-of-the-art solutions for such kind of graph-based keyword and search word extraction [59] are based on extensions of the well-known algorithms PageRank [92] and HITS [55]. As a prerequisite, it is necessary to explain, how those semantic term relations can be extracted from texts which can then be used to construct term graphs (word nets) that are analysed by these solutions.

4.4 Measuring Word Relatedness

Semantic connections of terms/words come in three flavours:

1. synonymy,
2. similarity and
3. relatedness.

In case of synonymy, two words are semantically connected because they share a meaning. Their semantic distance—to quantify this relation—is 0. However, words can be semantically connected, yet do not share a meaning. In this case, the semantic distance is greater than 0 and can reflect either similarity and/or relatedness of the words involved. As an example, 'cat' and 'animal' are similar and related. However, 'teacher' and 'school' are related, but not similar.

In the simplest case, a lexical resource like Roget's Thesaurus [108] is available as a knowledge base which makes it possible to directly check for the words' synonymy. However, the task is getting more difficult when only text corpora or standard dictionaries are at hand. In these cases, synonymy of words cannot be directly derived or even taken for granted. Here, measures to quantify the semantic distance between words can be applied. A very low distance is often a sign for word synonymy, especially if the two words in question often appear together with the same words (have the same neighbours). Recently, a very effective method [128] for synonym detection using topic-sensitive random walks on semantic graphs induced by Wikipedia and Wiktionary has been introduced. This shows, that such tasks can be carried out with high accuracy even when static thesauri or dictionaries are unavailable.

Returning to measures to detect and assess word relatedness, the frequency of the co-occurrence of two terms/words in any order in close proximity in a text or text corpus is a first reliable indication for an existing semantic relatedness. Two words w_i and w_j are called *co-occurrents*, if they appear together in close proximity in a text document D. In contrast to collocations [100] and n-grams [64], which refer to contiguous sequences of a given number of textual units such as words, the order in which these words co-occur is not considered.

The most prominent kinds of such co-occurrences are word pairs that appear as immediate neighbours or together in a sentence. Words that co-occur with a higher probability than expected stand in a so-called syntagmatic relation [25] to each other. Those word pairs are also known as significant co-occurrences and are especially interesting as the words they comprise are usually semantically related (they did not co-occur by chance). The technique of statistical co-occurrence analysis is a popular means to detect such statistically significant co-occurrences as it is very effective and can be easily implemented.

A weight function $g((w_i, w_j))$ indicates, how significant the respective co-occurrence is in a text. If the significance value is greater than a pre-set threshold, the co-occurrence can be regarded as significant. Rather simple co-occurrence measures are, e.g. the frequency count of co-occurring words and the Dice [28] similarity coefficient DSC (see formula 4.1) and the similar Jaccard [48] index.

$$DSC(w_i, w_j) = \frac{2 \times |S_{w_i} \cap S_{w_j}|}{|S_{w_i}| + |S_{w_j}|} \tag{4.1}$$

For DSC, S_{w_i} and S_{w_j} denote the sets of sentences containing word w_i or word w_j respectively, $|S_{w_i}|$ is the number of sentences word w_i occurred in and $|S_{w_i} \cap S_{w_j}|$ is the number of times word w_i and w_j co-occurred on sentence level.

More advanced formulae rely on the expectation that the appearance of two words in close proximity is statistically independent (a usually inadequate hypothesis). With this hypothesis, however, the deviations between the number of observed and expected co-occurrences in real text corpora can be calculated. Therefore, a significant deviation leads to a high significance value. Co-occurrence measures based on this hypothesis are, for instance, the mutual information measure [18], the Poisson collocation measure [100] and the log-likelihood ratio [29]. Stimulus-response experiments have shown that co-occurrences found to be significant by these measures correlate well with term associations by humans [46].

All these presented statistical and corpus-based approaches can, however, only be used to obtain unspecific relationships of the co-occurrents involved as it is only possible to make a statement in the form of ‘word A has something to do with word B’ (and vice versa) about them. For applications such as query expansion or the grouping of terms, this level of specifity is likely sufficient, for more advanced tasks e.g. the realisation of a question answering system, more specific semantic relationships need to be extracted.

In order to address this problem, in recent years, several graph- and knowledge-based distance measures [17, 104] have been developed that make use of external resources such as the manually created semantic network WordNet [84],[1] a large lexical database containing semantic relationships for the English language that covers relations like polysemy, synonymy, antonymy, hypernymy and hyponymy (i.e. more general and more specific concepts), as well as part-of-relationships. These measures apply shortest path algorithms or take into account the depth of the least common subsumer concept (LCS) to determine the semantic distance between two given input terms or concepts. With the help of these resources, it is instantly possible to determine their specific semantic relationship as well.

It is also common to measure the degree of relatedness of two terms/words w_i and w_j by

1. determining their respective k most significant co-occurrents,
2. treating the set of those terms as their (semantic) global contexts and
3. computing the similarity of these contexts.

Technically, these contexts are realised as term vectors following the vector space model [110] known from the field of information retrieval. A paradigmatic relation [25] between two words can be derived if their context vectors have a significant amount of entries in common. The similarity of two term vectors $\overrightarrow{t_1}$ and $\overrightarrow{t_2}$ can be determined using the cosine similarity measure

$$cos(\overrightarrow{t_1}, \overrightarrow{t_2}) = \frac{\sum_{i=1}^{n} t_{1i} \times t_{2i}}{\sqrt{\sum_{i=1}^{n} t_{1i}^2} \times \sqrt{\sum_{i=1}^{n} t_{2i}^2}} \tag{4.2}$$

or by calculating the overlap of term vectors, e.g. using the Dice coefficient [28] as presented in formula 4.1. The commonly used Euclidean distance and the Manhattan distance (taxicab distance) are further examples to measure the closeness of term vectors at low computational costs. This way, it is possible to determine a relatedness between terms even if they do not co-occur e.g. on sentence level. The equivalence class of semantically similar words obtained this way is known as a specific concept that can represent them. In this sense, a concept is a (rather constant) topical abstraction for concrete terms. However, it is still possible that a concept's meaning is changing over time. This is reflected in a volatility of the term vectors involved [45] which can be measured.

The recently developed approach of so-called word embeddings is another way to derive the context of words/terms using vectors [82]. Word embeddings are vectors

[1]For the German language, the analogous resource is GermaNet (http://www.sfs.uni-tuebingen.de/GermaNet/).

to numerically represent words which are learned using a skip-gram recurrent neural net architecture provided by the tool word2vec (https://code.google.com/archive/p/word2vec/) made publicly available by Google. Vectors of similar words are placed next to each other in the used vector space consisting of typically several hundred dimensions (to be specified). This makes it possible to easily list semantically related words close-by. Also, these vectors can be used to compare words regarding their similarity. This comparison can simply be carried out by e.g. applying the cosine similarity measure. As an additional feature, these vectors also encode linguistic regularities and patterns. The vectors' usefulness has been demonstrated for numerous applications such as the topical grouping of words.

In any case, the resulting term-term-matrix contains the (usually normalised) determined distance or similarity values for all term combinations in the text or text corpus, whereby—in case of determined similarities—values near 0 indicate low (small overlap of context vectors) and values near 1 high (large overlap of their context vectors) similarity. With the help of these values, it is also possible to determine the mean term-term-similarity inside a cluster of terms in order to evaluate its topical homogeneity, too.

4.4.1 Co-occurrence Windows of Varying Sizes

As mentioned, classic approaches for co-occurrence analysis count the occurrence of word pairs that appear as immediate neighbours or occur on sentence level. This lack of flexibility can be addressed by using a co-occurrence window of arbitrary size represented by the neighbourhood τ ($\tau \geq 1$) which determines the number of words (and therefore their positions) before and after a word of interest (punctuation marks will be ignored) that will be considered co-occurrents. As an example, for the phrase 'The quick brown fox jumps over the lazy dog.', the word of interest being 'fox' and $\tau = 2$ the list of co-occurrents would contain the words 'quick', 'brown', 'fox', 'jumps', 'over' from which all term combinations/co-occurrences (e.g. 'quick brown', 'quick fox', 'quick jumps' and so on) can be determined. The size of the co-occurrence window will therefore be $2 \times \tau + 1$. To evaluate the correlation between two words A and B from the set of all words T_C in a text corpus C, the number of times a specific co-occurrence is found, must be counted. In formula 4.3, function $f_\tau(A, B)$ therefore returns the number of times the words A and B with their respective positions P_A and P_B in the text co-occurred given τ:

$$f_\tau(A, B) := |\{B \in T_C|\ \exists A \in T_C : B \in K(A, \tau)\}| \qquad (4.3)$$

The environment $K(A, \tau)$ is defined by formula 4.4:

$$K(A, \tau) := \{W \in T_C|\ |P_A - P_W| \leq \tau\} \qquad (4.4)$$

The following applies $A \in K(B, \tau) \Leftrightarrow B \in K(A, \tau)$ and $f_\tau(A, B) = f_\tau(B, A)$, too.

Then, the correlation of two words A and B can be calculated as follows:

$$Corr_\tau(A, B) := \frac{f_\tau(A, B)}{\max_{t \in T_C} f(t)} \tag{4.5}$$

Here, $\max_{t \in T_C} f(t)$ denotes the frequency of the most frequent word t in C.

For all words A and B, the correlation value can be easily calculated this way. Afterwards, a ranking should be applied in order to obtain a list of co-occurrences that is ordered by the correlation values whereby each co-occurrence is assigned a rank according to its correlation value. The co-occurrence with the lowest possible rank (1) has the highest correlation value. Co-occurrences with the same correlation value will receive the same rank. This makes it possible to compare the rankings of different co-occurrence measures that usually return significance values of incomparable value ranges. However, for this purpose, these co-occurrences must be ranked in the same way, too.

In order to show the validity of the correlation approach presented, several experiments have been conducted. Significant co-occurrences obtained from the renowned log-likelihood measure and the equally important Poisson collocation measure are generally considered to be relevant. Therefore, the idea is to compare their ranks with the ranks obtained from the correlation measure. Here, the goal was to show that the ranks from the three co-occurrence measures do not differ greatly. As an example, the rank of a specific co-occurrence 'car, BMW' might be 1 according to the log-likelihood measure, yet, its rank according to the correlation might be 3. The difference (distance) of these two values is 2. These distances have been calculated for all co-occurrences determined in the experiments. In Figs. 4.1 and 4.2, the average of those rank differences for 10 corpora each consisting of 10 topically homogeneous texts from the German news magazine 'Der Spiegel' is depicted for the first 120 co-occurrences. To make the results more meaningful (in these tests, the log-likelihood measure and the Poisson collocation measure determine co-occurrences on sentence level) the parameter τ for the correlation measure has been set to 8. The co-occurrence window size is therefore 17 (the window acts as an artificial sentence), which is a good estimation for the average number of words per sentence [24] for the English language.

It is easy to see, that especially among the first 50 most important co-occurrences found by the log-likelihood measure, the correlation measure regards those co-occurrences as significant, too. Their rank values do not differ much from the values assigned by the log-likelihood measure.

For the less significant co-occurrences, however, these ranks differ increasingly. This is not surprising as the probability that less frequently co-occurring words receive the same rank using different measures decreases. The same applies to the comparison of the Poisson collocation measure with the correlation measure. Also, although not depicted here, the curves for different values of τ such as 6, 7, 9 and 10

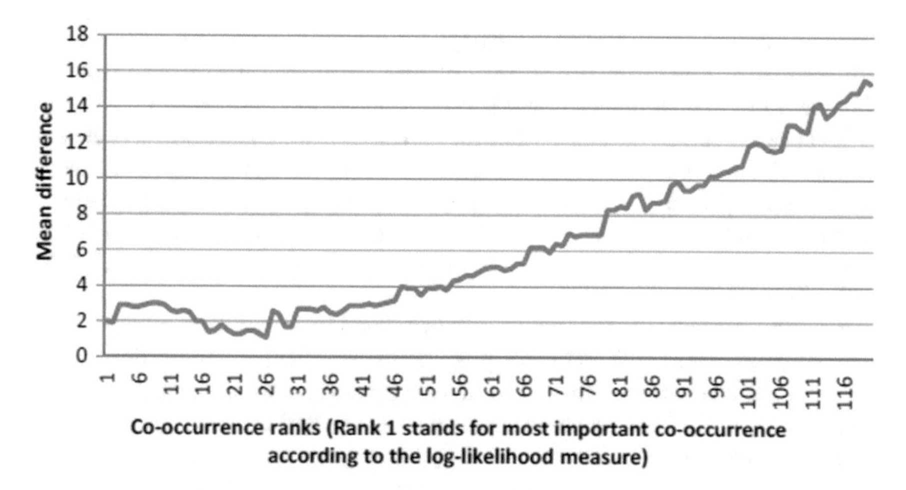

Fig. 4.1 Average of rank differences (log-likelihood measure vs. correlation measure)

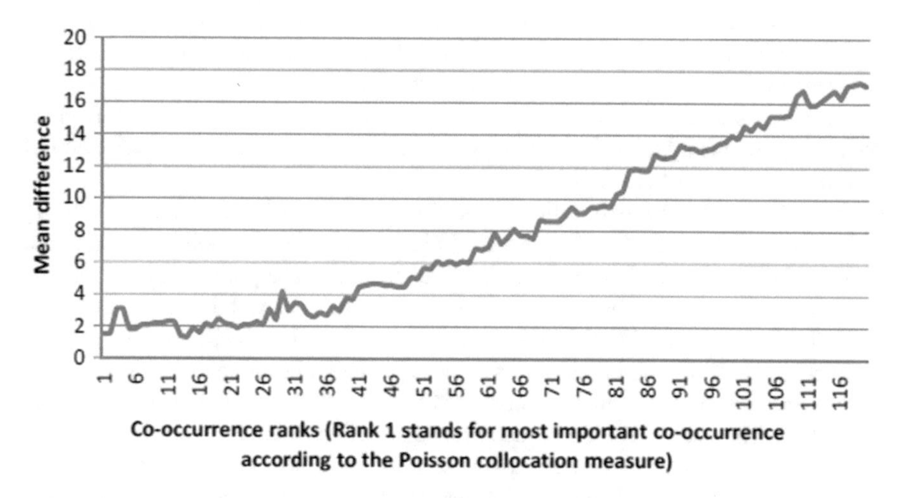

Fig. 4.2 Average of rank differences (Poisson colloc. measure vs. correlation measure)

look similar. This shows that the same co-occurrences are regarded as significant by all three measures.

4.4.2 N-Term Co-occurrences

So far, only co-occurrences consisting of two terms/words have been considered. The notion of 'n-term co-occurrence' addresses this restriction and can be defined as an unordered set of n terms $n \geq 2$ appearing together within a window of specified size, on sentences level or within paragraphs. A σ-significant n-term co-occurrence requires that its words appears at least σ times together in the selected word environment in order to be taken into account for further considerations. Although co-occurrences containing any parts of speech can be detected this way, it is sensible to only take into account nouns and names as they represent terms that carry a meaning, a characteristic especially useful for applications such as finding information in large text corpora or in the WWW. This is why the author selected the name 'n-term co-occurrence' and not 'n-word co-occurrence' or 'multi-word collocation' which can be found in literature [100], too.

Also, this notion's meaning must be distinguished from the meaning of a 'higher-order co-occurrence' [12] that usually represents so-called paradigmatic relations [25] of word forms that in turn can be derived by comparing the semantic contexts (e.g. vectors of significantly co-occurring words) of word forms. Higher-order co-occurrences cannot be directly extracted by parsing the mentioned text fragments. N-term co-occurrences, however, represent important syntagmatic relations between two or more word forms and can be easily extracted in the process of parsing text. Significant n-term co-occurrences can also be found using the approaches presented in the previous section to determine and rank 2-term co-occurrences. As 2-term co-occurrences can be used for query expansion purposes, n-term co-occurrences might be an even more effective means to filter out documents in search processes. In order to investigate this hypothesis, it is sensible to find out what a typical distribution of n-term co-occurrences looks like?

In order to answer this question, a set of experiments has been conducted using different corpora with topically clustered articles on 'German politics', 'cars' and 'humans' of the German news magazine 'Der Spiegel'. The politics corpus consists of 45984 sentences, the corpus on cars has 21636 sentences and the corpus on topics related to humans contains 10508 sentences. To conduct the experiments, stop words have been removed, only nouns and names as the only allowed parts of speech have been extracted and base form reduction has been applied prior to the co-occurrence extraction. The value τ for the co-occurrence neighbourhood has been set to 8 as in the previous section. The n-term co-occurrences must have appeared at least twice in the corpus (in two different τ neighbourhoods) in order to be called significant. In Fig. 4.3, the distributions of 2-, 3-, 4-, 5-, 6-, 7-, and 8-term co-occurrences for the corpus on 'German politics' are presented (logarithmic scale on X-axis). The distributions for the other two corpora mentioned look alike.

It can be clearly seen, that the most frequent co-occurrence is a 2-term co-occurrence and was found 476 times. This specific co-occurrence ('Euro, Milliarde') appeared more than three times more often than the most frequent 3-term co-occurrence ('Euro, Jahr, Milliarde') that was found 130 times. This observation is interesting as it shows that by taking into account just one additional term (in this case 'Jahr', German for 'year') the semantic context of a 2-term co-occurrence can be specified much more precisely. This is particularly true for co-occurrences that appear very frequently and therefore in many different semantic contexts in the used corpus.

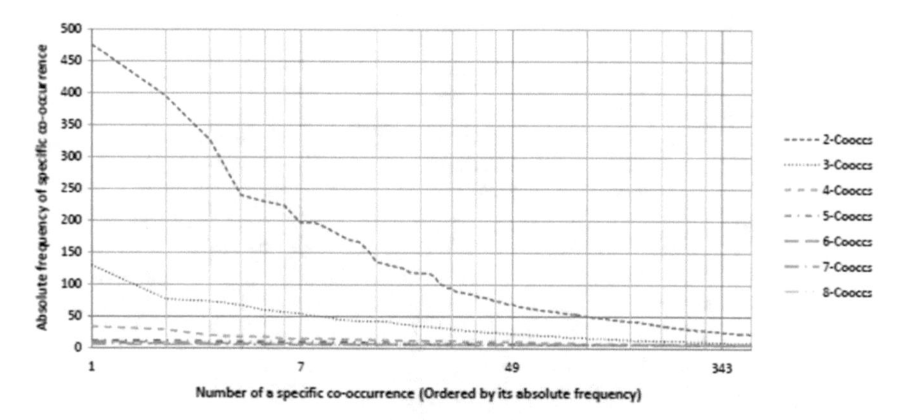

Fig. 4.3 Distribution of n-term co-occurrences (logarithmic scale on X-axis)

By adding another co-occurring term such as 'Bund' to the most frequent 3-term co-occurrence, the absolute frequency of this 4-term co-occurrence is again drastically reduced to 12. However, the frequency of specific 5-term, 6-term, 7-term or 8-term co-occurrences is generally much lower than the frequency of 2-term, 3-term or 4-term co-occurrences. When using five or more terms in a query, it is therefore likely that only a very low number of results will be returned by the requested search system, or even no results at all. However, the computational effort to determine such types of n-term co-occurrences will rise significantly due the increased number of possible term combinations. Even so, for the effective refinement of queries, 3-term or 4-term co-occurrences contain—as assumed in the hypothesis—valuable expansion terms that can be used to effectively filter out documents.

4.4.3 Finding Term Associations

As already indicated, while the classic co-occurrence measures mentioned return the same significance value for the relation of a term A with another term B and vice versa, an undirected relation of such kind often does not represent real-life relationships very well. As an example, one might instantly and strongly associate the term 'BMW' with the term 'car', but not vice versa. The strength of the association of 'car' with 'BMW' might be much lower. Thus, it is sensible to deal with directed and therefore more specific term relations instead. To determine the strength of such an association of term A with term B on sentence level in a text corpus C, the author proposes the usage of formula 4.6 as a measure of confidence known from the field of association rule mining.

$$Assn(A \rightarrow B) = \frac{|S_A \cap S_B|}{|S_A|} \tag{4.6}$$

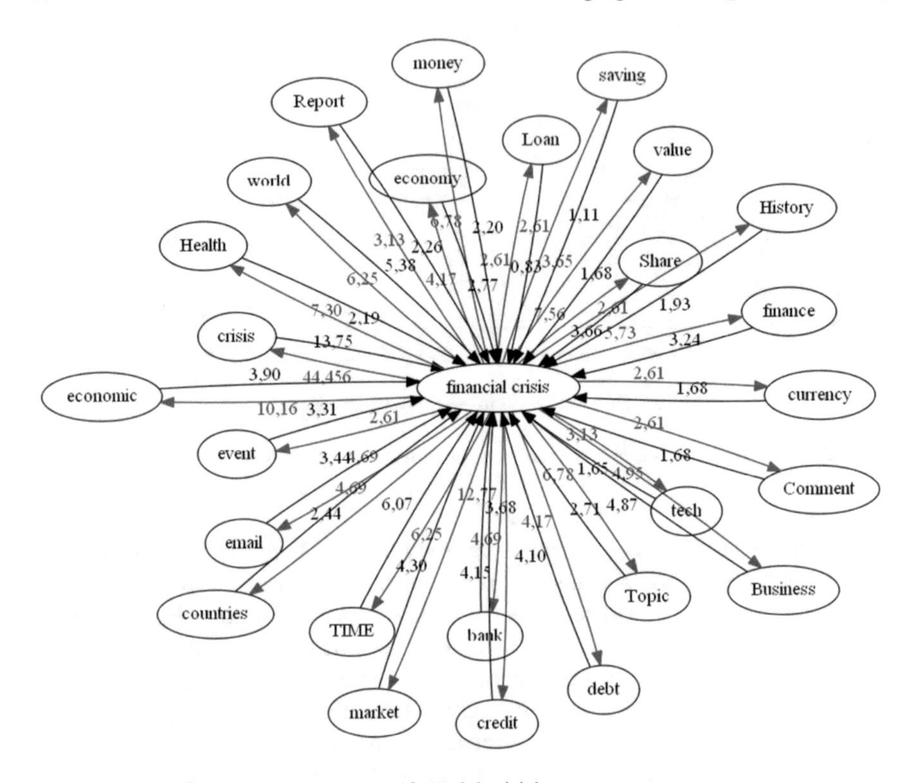

Fig. 4.4 Association graph for the query 'financial crisis'

Here, S_A and S_B denote the sets of sentences containing term A or term B respectively, $|S_A|$ is the number of sentences term A occurred in and $|S_A \cap S_B|$ is the number of times term A and B co-occurred on sentence level.

A relation of term A with term B obtained this way can be interpreted as a recommendation of A for B when the association strength is high. Relations gained by this means are more specific than undirected ones between terms because of their added direction. They resemble hyperlinks on websites. In this case, however, such a link has not been manually and explicitly set and carries an additional weight that indicates the strength of the term association.

The set of all such relations obtained from a text or text corpus also represents a directed co-occurrence graph, also called association graph (as seen in Fig. 4.4), that can be analysed by e.g. the extended HITS algorithm [59] to find the main topics and their sources (important inherent, influential aspects/basics). In the next section, a method will be introduced to use them as search words in order to find similar and related web documents and to track topics as well.

Another way to construct directed term graphs is the usage of dependency parsers [79]. After a part-of-speech tagging is applied, they identify syntactic relationships among words in the text and generate dependency trees of them for each sentence.

Detecting term relations using lexico-syntactic patterns is another well-known approach [106] for this task. Here, interesting patterns of parts of speech and/or word forms in a specified order are defined and searched for in the texts. This way, part-of- and is-a-relationships can be detected easily. A pattern like '[NN] like [NN] and [NN]' can be used to uncover hyponyms (subsumable concepts) and hypernyms (superordinates) and, thus, determine the direction of the relation between the terms referred to.

4.5 Measuring Text Relatedness

The most common approach to determine the semantic similarity or distance of text documents is by applying the vector space model. Here, each document is represented by its document vector which contains the text's characteristic terms along with their scores (typically, a TF-IDF-based statistic [8] is used) for their importance/significance. The document vectors act as a 'bag of words' that neglect the order or sequence in which their terms originally occurred in order to simplify subsequent computations. The main problem with this approach is that no semantic relations between the terms is encoded in them, even though with a high probability these relations exist. Every single term found in a text corpus is globally assigned one component (and position) in the vectors. This causes further problems: in the vector space, synonyms will constitute different dimensions (instead of one) and homonyms will be mapped to one and the same dimension despite their semantic differences.

Even so, the cosine similarity measure can be applied to compare the document vectors of all given document pairs and return their similarity score. The idea behind this approach is that similar documents contain with a high probability the same terms. This approach is easy to implement and effective. However, due to the high-dimensional space consisting of potentially many thousands of terms, the individual document vectors contain the same number of components with many of them being weighted with 0 (they do not appear in the particular document at all). The term-document matrix constructed from these vectors indicates which term appears in which documents. Thus, it is both typically sparse and high-dimensional.

In some cases, standard distance or similarity measures based on this bag-of-words model do not work correctly (with respect to human judgement), mostly if different authors write about the same topic but use a different wording for doing so (vocabulary mismatch). The overlap of the respective document vectors would be small or even empty. The reason for this circumstance can be seen in the isolated view of the words found in the documents to be compared without including any relation to the vocabulary of other, context-related documents.

Moreover, short texts as often found in posts in online social networks or short (web) search queries contain only a low number of descriptive terms can therefore often not be correctly classified or disambiguated. Another disadvantage of these measures is that they cannot find abstractions or generalising terms by just analysing the textual data provided. For this purpose, static lexical databases such as the already mentioned WordNet [84] must be consulted as a reference, e.g. to uncover synonymy relations. Despite their usefulness, these resources are—in contrast to the human brain—not able to learn about new concepts and their relationships on their own.

The bag-of-concepts model has been conceived to address the shortcomings of the bag-of-words model. The idea is to combine semantically related terms in topical clusters, called concepts. Instead of individual document-specific features, the document vectors indicate whether particular concepts are covered in the documents. The mapping from individual features to concepts, however, means additional computational overhead. It is usually carried out by means of external resources such as Wikipedia or WordNet that provide possible labels for concepts as well as exploitable relations between individual terms. An example for this approach is called Explicit Semantic Analysis (ESA) [35]. In this solution, a term is represented by a vector of weighted Wikipedia concepts (articles) that it is mentioned in. By relying on concepts, it is possible to abstract from specific features when comparing documents semantically as their general topical relatedness is determined. Further approaches to group terms in meaningful concepts are given in the next section. Nevertheless, the authors of this article also introduced an extended cosine similarity measure which takes into account the similarity of individual terms (following the bag-of-words model) or extracted concepts (following the bag-of-concepts model) when comparing the similarity of two document vectors. A similar approach is presented in [40]. Here, the authors measure the relatedness of texts by first determining relatedness and specifity scores of the documents' words and second by including them into a particular document relatedness calculation.

The previously mentioned approach of word embeddings has been successfully applied to derive semantic distances of documents [72], too. Word embeddings have also been successfully used to determine the similarity between short texts [54]. In this context, a relevant extension to word embeddings has been presented in [73]. The authors propose the computaton of distributed representations (paragraph vectors) for sentences, paragraphs and whole documents. It is possible to use these representations for information retrieval and sentiment analysis tasks. In [93], an unsupervised model (called Sent2Vec) for the composition of sentence embeddings using word vectors in combination with n-gram embeddings is introduced. The authors show that this model achieves state-of-the-art performance on sentence similarity tasks.

In this regard, it is noteworthy that in recent years many researchers were attracted by the SemEval (Semantic Evaluation) international workshops and their task of determining semantic textual similarity (STS) of cross-lingual and monolingual pairs of sentences. Thus, a large variety of approaches and systems for this purpose have been proposed [1]. In this book, a graph-based technique will be introduced that addresses this problem in a new manner.

In any case, the resulting document-document matrix contains the (usually normalised) determined distance or similarity values for all document combinations in the text corpus.

4.6 Grouping Words and Texts

In many text mining applications, the (usually unsupervised) topical grouping (clustering) of objects such as texts or terms/words contained in them are common tasks. Generally speaking, the usual goal in this setting is to

1. compute topically homogeneous clusters of semantically close objects
2. and to achieve heterogeneity between different clusters (semantically distant objects should be put into different clusters).

However, it is—besides semantic considerations and depending on the task at hand—also possible to cluster those objects with regards to other aspects, e.g. the same parts of speech or pre- and suffixes. The most common application of clustering techniques in text mining is to semantically group text documents. In contrast to text classification, there are no pre-defined categories to assign those texts to. They are clustered based on their inherent properties and features. For this purpose, it is necessary to determine the texts' semantic relatedness as described before.

Also, it is often necessary to group terms/words according their topical relatedness. As mentioned above, the well-known method called Latent Semantic Indexing (LSI) [27] can be used to automatically group similar terms found in document collections by mapping synonymous and related terms to the same topical dimension or concept. It reduces the dimensionality of the typically sparse original term-document-matrix using the algebraic technique called Singular Value Decomposition (SVD). Naturally, a problem is to interpret the gained concepts to which the terms are assigned.

Unsupervised topic modelling techniques such as Latent Dirichlet Allocation (LDA [13]) have been successfully applied to group semantically related terms as well. This technique tries to infer word clusters from a set of documents based on the assumption that words from the same topic are likely to appear next to each other and therefore share a related meaning. Here, deep and computationally expensive (hyper)parameter estimations are carried out and for each word, the probability to belong to specific topic is computed in order to create those constructions. Despite this drawback, a major shortcoming of this approach is that the number of topics (clusters) to be generated (which are to be filled with terms) needs to fixed as an input parameter.

This is a too serious restriction for important application areas such as general recommender systems, because it necessitates visualisation of the underlying datasets and intervention by human experts and further prohibits recommendations to be offered on a continuous basis and in an unattended mode. This problem is particularly relevant for the domain of automatic text analysis. For instance, a book such as conference proceedings could cover a variety of major and minor topics with each one having its own domain-specific terms. It would be—even depending on the granularity required—hard to manually estimate the correct number of such topics.

4.6.1 Non-hierarchical (Flat) Clustering

While an abundance of clustering algorithms has been devised [32], the classical and most widely used one is k-Means [77]. This algorithm is an example for a non-hierarchical clustering technique. The number of clusters to be generated has to be specified as an input parameter k. Initially, k elements will be picked as representatives ('means') of the different clusters. In the following steps, the other data elements (in this case documents or terms) are associated with their closest means (here, the computed scores in the document-document-matrix or term-term-matrix), and new cluster centroids are calculated repeatedly until the centroids do not change anymore and convergence is reached. This way, a local optimum can be achieved. In order to increase the probability of finding a global optimum, k-Means is executed several times with different start configurations.

Its major disadvantage—as in the case of LDA—is that the number of clusters k to be generated needs to be set before the algorithm runs which—depending on the goal at hand—is hard to determine properly. The advantage of not having to estimate this parameter is especially beneficial when information on the heterogeneity or homogeneity of data objects is insufficient.

Another algorithm for flat clustering, which—in contrast to k-Means and LDA—does not require a pre-set number of clusters/topics to be generated, is the so-called Chinese Whispers algorithm [11] which employs a simple, yet effective technique for label assignment in undirected graphs. First, a specific label is assigned to each vertex. In the next steps, the vertices determine and take on the most popular label in their direct neighbourhood. After the algorithm's execution, those nodes marked with the same label belong to one same cluster.

This algorithm was successfully used in the field of natural language problems, especially to semantically group terms in text corpora and single texts (see Fig. 4.5) but also to determine the correct part of speech of words. It is usually applied on undirected semantic graphs that contain statistically significant term relations found in texts.

4.6.2 Hierarchical Clustering

In contrast to non-hierarchical methods, hierarchical clustering techniques determine a hierarchy of element groups in each step and work in an agglomerative (bottom-up) or divisive (top-down) way [46]. In the first case, initially all data elements to be clustered reside in their own clusters. In subsequent steps, two clusters with the highest similarity are merged. In the second case, initially all data elements reside in a single (possibly big) cluster. In subsequent steps, the cluster with the least degree of coherence is split. In both cases, the clustering process is stopped when a reasonable compromise between a small number of clusters and a high degree of homogeneity of the data elements inside the clusters has been reached. Also, in both

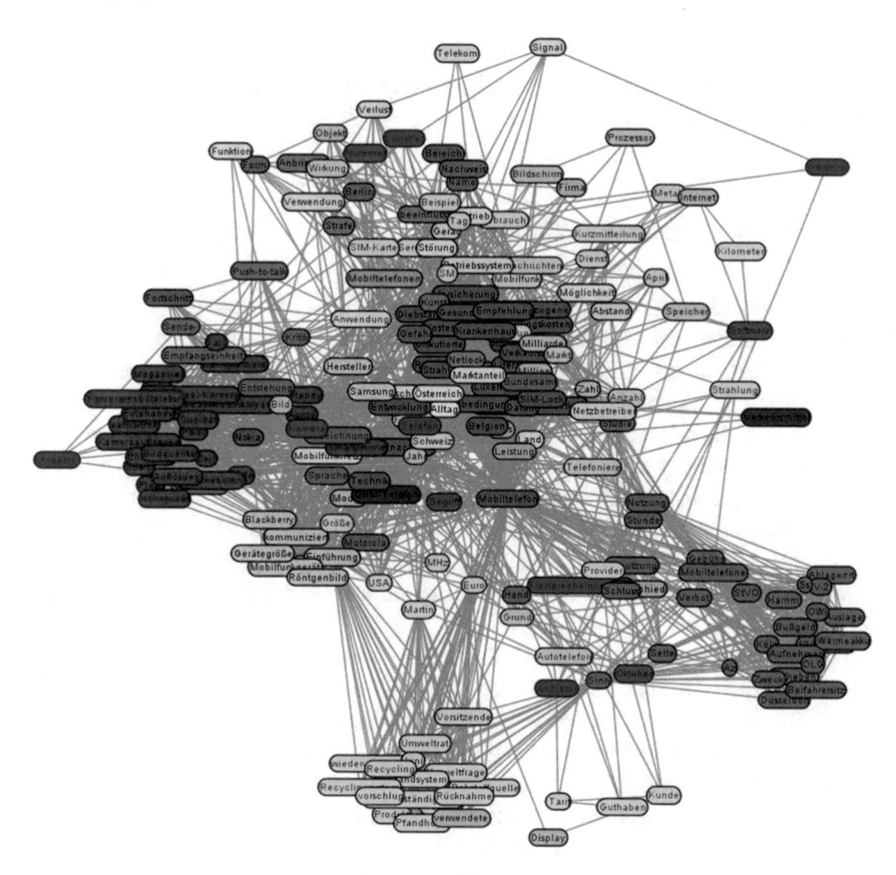

Fig. 4.5 Clusters of terms from the German Wikipedia-article 'Mobiltelefon' generated by Chinese Whispers

cases, methods are needed to determine the similarity of two complex clusters. The following approaches can be used for this purpose:

- Single-link: This approach generates a new cluster of the two clusters, whose elements have the highest similarity. It tends to lead to large clusters.
- Complete-link: This approach determines the maximum distance between two clusters and merges the clusters with the lowest maximum distance. It tends to lead to small clusters.
- Average-link: This approach calculates the average distance between the objects in two clusters. The clusters with the lowest average distance are merged. In general, this leads to clusters of almost the same size.

Hierarchical clustering algorithms—both agglomerative and divisive ones—are able to dynamically determine the number of clusters modelling a given dataset. It suffers, however, from another drawback impairing its applicability for many practical purposes, viz. that a set of objects to be clustered must be available for processing in its entirety. In contrast to this, the objects considered by recommender systems are added one by one, and such systems are expected to be operational all the time and permanently available as web services.

4.6.3 *Other Clustering Techniques*

The following algorithms can be used to cluster graphs of documents or terms that reflect and encode the relatedness of the respective entities. In 2003, Flake et al. [33] introduced a cut-clustering algorithm to determine a set of minimum s-t-cuts in undirected graphs with edge weights. First, the original graph is augmented by inserting an artificially inserted sink, which is linked to any other vertex in the graph with the edge weight α. A cut tree from this augmented graph is then computed, and the sink is removed afterwards. This leads to a separation of the cut tree into several connected components which are returned as clusters. Thereby, the clusters are weakly connected to the rest of the graph, only, because they are induced by minimum cuts. The inserted edges guarantee an expansion based on α within the clusters.

Spectral graph partitioning algorithms rely on calculating the eigenvectors of a graph's Laplacian or its adjacency matrix [81]. The method introduced by Capoccia et al. [19] also works on directed weighted graphs. Another algorithm of this class was presented by Qiu and Hancock [99]. It relies on the calculation of the Fiedler vector which partitions a given graph into two components. Random-walk-based algorithms rely on the intuitive idea to find dense subsets of vertices (clusters) of a graph that a random walker is not likely to leave [122]. This means that the members of such a cluster should be visited with a higher probability from within the cluster than from outside.

4.7 Summary

This chapter discussed basic problems in natural language processing and explained several ways to extract keywords from texts. Furthermore, a large variety of state-of-the-art text mining techniques have been discussed. Especially, graph-based methods to measure the semantic relatedness between words and texts has been elaborated on in detail while focussing on approaches that rely on the calculation and utilisation of co-occurrences and term associations. The understanding of these approaches is necessary to follow the considerations in the subsequent chapters. A detailed discussion on existing solutions for clustering terms and texts that can be applied in manifold text mining tasks concluded the chapter. Their shortcomings motivate the concepts for decentralised library management as well as their realisation which will both be covered in Chap. 7.

Chapter 5
Local Librarians—Approaches for Context-Based Web Search

5.1 Local Search Support

The idea to conceive and develop 'local librarians' in form of interactive search tools was born out of the described shortcomings of how current web search engines handle queries and present search results. It was found that an automatic recommendation of search words and queries that regards the context of the search subject and user interests would be of great help to the user. In order to be able to do so, these tools analyse local documents in order to provide a more precise description of the users' information needs in form of appropriate keywords for the current search context [59] and do not transfer any user-related information to the chosen (centralised) web search engine. The search context originates on the user's computer, not after the search engine has already been contacted.

The basic conceived software architecture is depicted in Fig. 5.1. It is to be seen that the unrivalled huge and well-indexed databases as well as access mechanisms of the big web search engines are combined with a local pre-processing agent (the local librarian) or software. This agent has access to the local (and, with the user's agreement maybe confidential) files of the user and may even establish a fine-granular user profile. As this data is kept local, there is no danger for privacy and security. Current search queries can then be enriched by the knowledge from these local files and previous search sessions in order to present keyword suggestions or query expansions to the user who can decide on whether they shall be sent as a query to the (remote) web search engine or not, which in turn returns its results in the known manner.

5.2 Query Expansion and Unsupervised Document Search

One of the first locally working methods co-developed by the author that can be easily applied in a local agent is to retrieve web documents (especially scientific articles) using automatically generated queries given an initial query Q and a local

M. Kubek, *Concepts and Methods for a Librarian of the Web*,
Studies in Big Data 62, https://doi.org/10.1007/978-3-030-23136-1_5

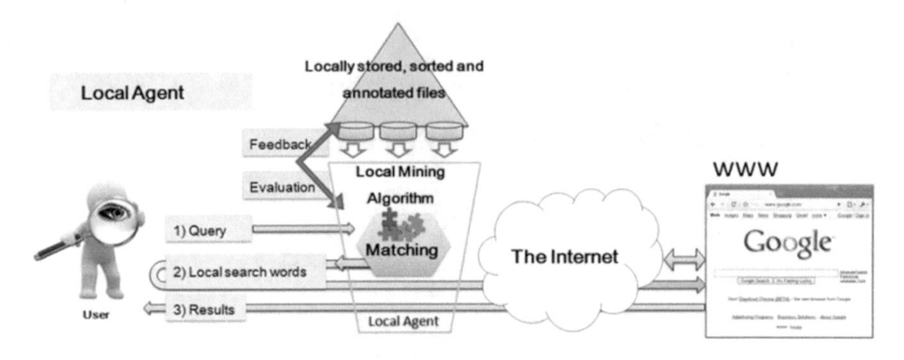

Fig. 5.1 Cooperation of the local librarian and web search engines

document set $D1$ [71]. The idea of using expansion terms was to address the vocabulary mismatch problem. A search for e.g. 'notebook' should also return results for its synonym 'laptop'. For this purpose, the set $D1$ is used to expand the initial query Q with at most two relevant keywords semantically connected to Q. The generated queries are then sent to a web search engine. Optionally, the 10 best results for each expanded query will be automatically downloaded and ranked according to $D1$. The method was conceived in the project 'Search for text documents in large distributed systems' supported by the German Research Foundation [130]. It is working for German and English text documents.[1]

This is how the implemented query expansion method works:

1. At first, an initial query Q and a document set $D1$ for query expansion are provided.
2. A profile P is then calculated using $D1$ by concatenating all texts in $D1$ to an aggregation text T and extracting the most significant k words of T. The significance is determined through the usage of a well-balanced reference corpus. With this corpus, it is possible to estimate the frequency of words in everyday-language. Words whose relative frequency in T significantly surpasses the one estimated from the reference corpus, will be ranked highly [136].
3. T is further used to calculate significant co-occurrences for each of the k words in P. These co-occurrences are saved in the co-occurrence graph G using a matrix.
4. A spreading activation algorithm [98] is then performed on this graph, which leads to an activation level and ranking for each of the k words in P related to the given query Q, whose elements are activated highly. The n highest ranked words in P are then considered the expansion set E.
5. The expanded queries $(Q_1 \ldots Q_m)$ are generated by using the activation levels (as a probability distribution) to determine the positions of the selected elements of E. Here, either one or two expansion terms/words have been added to Q.

[1] The applied software components for semantic text analysis were provided by the NLP department of the University of Leipzig (http://asv.informatik.uni-leipzig.de/en/).

Table 5.1 Average precision of the 10 first results

Number of expansion terms:	0	1	2
Average precision:	0.38	0.76	0.88

6. The expanded queries are sent to Yahoo! or Google.
7. (Optionally) The 10 best results for each query are downloaded and ranked while avoiding duplicates among the newly found documents. This is called the document set $D2$. To determine the ranking of the documents in $D2$, the profile P and the terms/words of the elements in $D2$ are considered as vectors and their cosine similarity is calculated.

In case that the initial query Q cannot be expanded with the document set $D1$, the reason might be that the topics in $D1$ differ too much among the documents or from Q itself. Then, only the initial query Q is sent to the web search engine. Therefore it is sensible to select documents with topics matching the query Q, because the algorithm will return more suitable expansion terms when topically homogeneous documents are provided.

This approach has been evaluated in a number of tests whereby only terms such as nouns (in their base form), proper nouns and names have been extracted from the document sets. For every initial query Q, the following tests have been conducted:

- web search without query expansion,
- web search with one expansion term and
- web search with two expansion terms.

Altogether, 150 initial queries from the topics geography, movie and tv, comics, history, chemistry and astronomy. The following table presents the achieved average precision for all queries generated (Table 5.1).

The results show that only one additional expansion term leads to a significantly higher number of documents which are similar to $D1$. The second expansion term improved this result once more, however to a lesser extent. The given query expansion strategy works best if the initial query contains an ambiguous word. It is also useful if it contains general words that often appear in the context of many other words. The expansion with their co-occurrents in $D1/P$ helps to make the respective query more precise.

In the given algorithm, nothing is said about how the initial query Q is determined. On the one hand, it can be an expression of a user's information need entered into an IR system. Additionally, (selected) documents in $D1$ (more precisely the elements in P) can be chosen as Q. This way, the algorithm's input would be restricted to $D1$ only and the result set $D2$ would depend on the automatically generated queries as well as the web search results. In this thinking, this approach is able to automatically create both initial and expanded queries in order to find similar documents in the WWW. In order to support the iterative nature of information search, it is sensible to add those documents in $D2$ to $D1$ that are actually topically related to Q and $D1$. This way, the newly acquired knowledge in $D2$ would be available as search context in future search sessions, too. While one could argue that a 'filter bubble' induced by $D1$ could be the consequence of this procedure, the document set is definitely of help for e.g. experts using a specific terminology to properly expand 'unusual' queries.

At the same time, this approach can be used to extensively search the web for a topic of interest. This bootstrapping strategy has been successfully used to periodically search the web for new and relevant scientific publications in an unsupervised manner as well as to download and archive them with respect to their topical orientation. In the next section, another approach is presented which can be used to automatically construct web queries in order to find similar and related web documents.

5.3 Finding Similar and Related Web Documents

Due to the large amount of information in the WWW and the lengthy and usually linearly ordered result lists of web search engines that do not indicate semantic relationships between their entries, the search for topically similar and related documents can become a tedious task. Especially, the process of formulating queries with proper terms representing specific information needs requires much effort from the user. This problem gets even bigger when the user's knowledge on a subject and its technical terms is not sufficient enough for doing so. Therefore, a solution is needed that can actively support the user in finding similar and related web documents.

It was found in [59] that the keywords and their sources (called source topics) of text documents obtained from the analysis of their directed co-occurrence graphs carried out by the extended HITS algorithm are well-suited to be used as search words for this purpose, too.

Moreover, topical sources of documents can be used to iteratively track topics they exhibit to their roots by e.g. using them as query terms which leads to further documents that in turn primarily deal with these topical sources. In order to gain a much clearer discriminability between the returned lists of main and source topics (less overlap is wished-for), an association should be taken into account only when its strength is greater or equal to the strength of the reverse association. The author therefore proposes formula 5.1 as an extension to formula 4.6 that accounts for this requirement.

$$Assn(A \rightarrow B) = \begin{cases} \frac{|S_A \cap S_B|}{\max_{t \in T_C} |S_t|}, & \text{if } \frac{|S_A \cap S_B|}{|S_A|} \geq \frac{|S_A \cap S_B|}{|S_B|} \\ 0, & \text{otherwise} \end{cases} \tag{5.1}$$

In addition to the previous expressions, $|S_B|$ is the number of sentences term B occurred in and $\max_{t \in T_C} |S_t|$ stands for the maximum number of sentences any (but a fixed term) term t from the set of all terms T_C in corpus C has occurred in. The resulting weight $\frac{|S_A \cap S_B|}{\max_{t \in T_C} |S_t|}$ of the directed link from A to B is derived by multiplying $\frac{|S_A \cap S_B|}{|S_A|}$ that indicates the basic association strength of A with B by $\frac{|S_A|}{\max_{t \in T_C} |S_t|}$ that accounts for the relative frequency of A (on sentence level) in the text as a measure for A's overall importance.

As mentioned before, a relation of term A with term B obtained this way can be interpreted as a recommendation of A for B when the association strength is high. The set of all such relations obtained from a text represents a directed co-occurrence graph ready to be analysed by the extended HITS algorithm that takes the term association values $Assn$ into account.

For this purpose, the formulae for the update rules of the HITS algorithm must be extended. The authority value of a node x can then be determined using formula 5.2:

$$a(x) = \sum_{v \to x} (h(v) \cdot Assn(v \to x)) \quad (5.2)$$

The hub value of a node x can be calculated using formula 5.3:

$$h(x) = \sum_{x \to w} (a(w) \cdot Assn(x \to w)) \quad (5.3)$$

The following steps are necessary to obtain two lists containing the analysed text's authorities and hubs based on these update rules:

1. Remove stop words and perform base form reduction on all word forms of interest (e.g. nouns/terms) in the text. (Optional)
2. Determine its directed co-occurrence graph G using formula 5.1 applied to all co-occurring term pairs.
3. Determine the authority value a(x) and the hub value h(x) iteratively for all nodes x in G using formulae 5.2 and 5.3 until convergence is reached (the calculated values do not change significantly in two consecutive iterations) or a fixed number of iterations has been executed.
4. Return all nodes in descending order by their authority and hub values with their representing terms and their authority and hub values.

These term lists (can be regarded as term clusters, too) can also show, nevertheless, an overlap when analysing directed co-occurrence graphs. Hence, this is a soft graph clustering technique. In each case, these lists are ordered according to their terms' centrality scores.

Table 5.2 presents for the article 'Financial crisis of 2007–08' from the English Wikipedia the two lists of main and source topics extracted by the extended HITS algorithm, whereby the following parameters have been used: removal of stop words, part-of-speech restriction to nouns and names, base form reduction and phrase detection.

The example shows that the extended HITS algorithm can determine clusters of the most characteristic terms (authorities) and source topics (hubs). Especially the list

Table 5.2 Terms with high authority (main topics) and hub values (source topics) from the Wikipedia-Article 'Financial crisis of 2007–08':

Term	Main topic score	Term	Source topic score
Bank	0.57	System	0.20
Crisis	0.42	Institution	0.19
US	0.34	Lending	0.19
Market	0.27	House	0.18
Mortgage	0.25	Loan	0.18
Credit	0.18	Risk	0.18
Institution	0.17	Market	0.17
House	0.16	Investment	0.17
Price	0.15	Mortgage	0.16
System	0.13	Credit	0.15

of source topics provides valuable insight into the analysed text's topical background which is useful to find related (not necessarily similar) content when used in queries. Another empirical finding was that the quality of the authority and hub lists improved when analysing clusters of semantically similar documents instead of single texts. The reason for this is that by using a larger textual basis for the analysis the calculated term association values are more statistically meaningful.

This approach for graph-based keyword and source topic extraction has been implemented in the service 'DocAnalyser' for the purpose of search word extraction and automatic query formulation in order to find similar and related web documents. This service will be introduced in Sect. 5.4.2.

5.4 Tools and Services

In this section, several tools and services that implement the foregoing algorithms shall be introduced.

5.4.1 The Local Search Agent 'FXResearcher'

The first example application called 'FXResearcher' that shall be presented here is an extension for the web browser Firefox. It implements (alongside others) the algorithm presented in Sect. 5.2.

This tool's aim is to support the user in searching for documents on the local computer and on web presences. For this purpose, it carries out a text analysis on a set of user-selected and local documents kept in specific user-defined directories

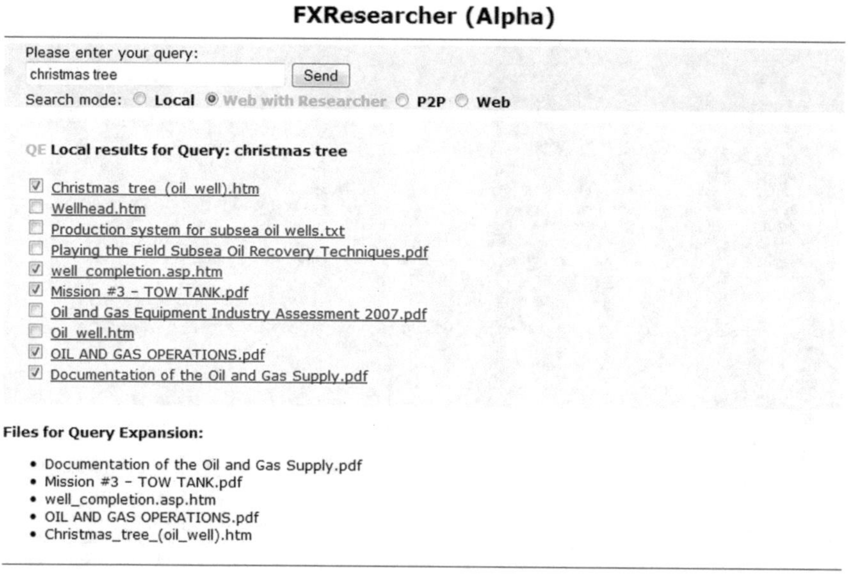

FXResearcher (Alpha)

Please enter your query:
christmas tree [Send]
Search mode: Local Web with Researcher P2P Web

QE **Local results for Query: christmas tree**

- [x] Christmas_tree_(oil_well).htm
- [ ] Wellhead.htm
- [ ] Production system for subsea oil wells.txt
- [ ] Playing the Field Subsea Oil Recovery Techniques.pdf
- [x] well_completion.asp.htm
- [x] Mission #3 - TOW TANK.pdf
- [ ] Oil and Gas Equipment Industry Assessment 2007.pdf
- [ ] Oil_well.htm
- [x] OIL AND GAS OPERATIONS.pdf
- [x] Documentation of the Oil and Gas Supply.pdf

Files for Query Expansion:

- Documentation of the Oil and Gas Supply.pdf
- Mission #3 - TOW TANK.pdf
- well_completion.asp.htm
- OIL AND GAS OPERATIONS.pdf
- Christmas_tree_(oil_well).htm

Fig. 5.2 FXResearcher: Local search results

in order to extract from them expansion terms to a given query for the next search iterations.

In addition, to account for user feedback, recently downloaded and evaluated documents may be used for result improvements in the next steps as well. Thus, this tool acts as a bridge between the local machine and the addressed web search engines.

In order to find matching local documents to be used for query expansion, FXResearcher integrates a full-text indexer based on Apache Lucene that incrementally indexes text documents in folders on the local computer that have been explicitly specified by the user. As seen in Fig. 5.2, documents in this index can then be interactively selected for query expansion as a form of relevance feedback when they contain terms of a search query.

This approach is promising, because a query will likely not be expanded with improper terms as it still occurs when query expansion will be first performed by the requested web search engine. The latter solution might completely fail when for instance an expert searches for documents with a quite specific term that does not occur in many documents. The usage of suggested improper but frequently occurring expansion terms could return only a few and inadequate results. The consideration of local knowledge for query expansion on the other hand can provide an expert with proper expansion terms on her or his area of expertise and therefore return more expected search results. Figure 5.3 depicts such a case: The query 'christmas tree' (in oil industry) is properly expanded. With this approach, even queries consisting of homonyms can be properly expanded.

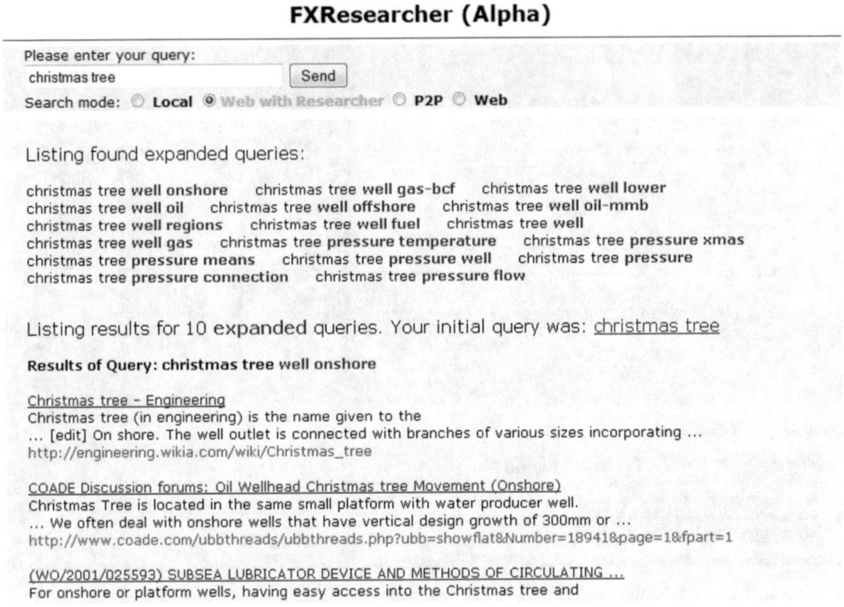

FXResearcher (Alpha)

Please enter your query:
christmas tree Send
Search mode: Local Web with Researcher P2P Web

Listing found expanded queries:

christmas tree well onshore christmas tree well gas-bcf christmas tree well lower
christmas tree well oil christmas tree well offshore christmas tree well oil-mmb
christmas tree well regions christmas tree well fuel christmas tree well
christmas tree well gas christmas tree pressure temperature christmas tree pressure xmas
christmas tree pressure means christmas tree pressure well christmas tree pressure
christmas tree pressure connection christmas tree pressure flow

Listing results for 10 expanded queries. Your initial query was: christmas tree

Results of Query: christmas tree well onshore

Christmas tree - Engineering
Christmas tree (in engineering) is the name given to the
... [edit] On shore. The well outlet is connected with branches of various sizes incorporating ...
http://engineering.wikia.com/wiki/Christmas_tree

COADE Discussion forums: Oil Wellhead Christmas tree Movement (Onshore)
Christmas Tree is located in the same small platform with water producer well.
... We often deal with onshore wells that have vertical design growth of 300mm or ...
http://www.coade.com/ubbthreads/ubbthreads.php?ubb=showflat&Number=18941&page=1&fpart=1

(WO/2001/025593) SUBSEA LUBRICATOR DEVICE AND METHODS OF CIRCULATING ...
For onshore or platform wells, having easy access into the Christmas tree and

Fig. 5.3 FXResearcher: Web search results after local query expansion

Furthermore, it was shown in [59] and mentioned above that the quality of web search results returned increased drastically when this approach of local query expansion is used for web search as the semantic context of the initial query could be specified more precisely this way.

5.4.2 'DocAnalyser'—Searching with Web Documents

As the second application, the interactive web-based application 'DocAnalyser' (http://www.docanalyser.de) is presented which provides solutions to the listed problems in Sect. 5.3. It enables users to find similar and related web documents based on automatic query formulation and state-of-the-art search word extraction. DocAnalyser uses the locally displayed contents of currently visited web documents as the only needed input to carry out this task. Additionally, this tool can be used to track topics across semantically connected web documents.

The user just needs to provide a web content to be analysed by clicking DocAnalyser's bookmarklet (downloadable JavaScript code in a web browser bookmark to send the selected web content to the DocAnalyser web service). This is usually the web page (or a selected part of it) currently viewed in the web browser (see Fig. 5.4).

DocAnalyser then extracts its main topics and their sources [58] (important inherent, influential aspects/basics) as presented in the previous section and automatically

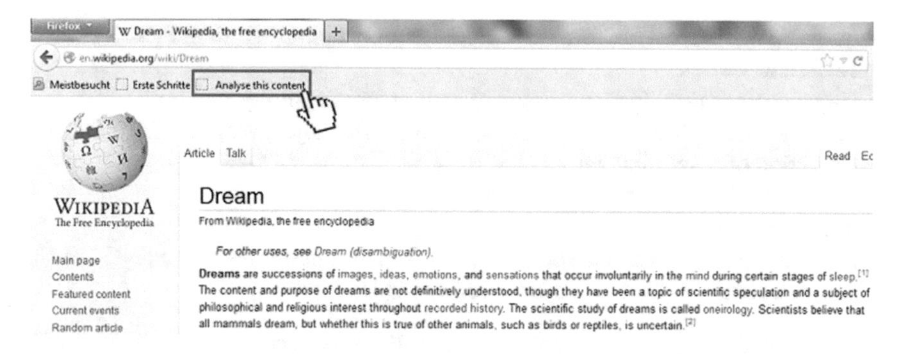

Fig. 5.4 DocAnalyser: Selection of web content to be analysed

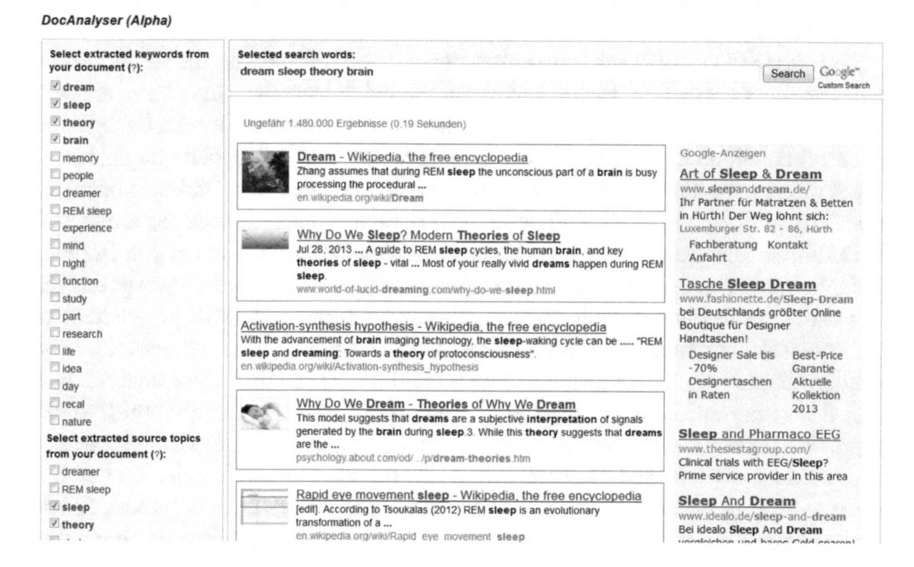

Fig. 5.5 DocAnalyser's result page

uses (at most four) of them as search words. The returned search words and web search results are generally of high quality. Usually, the currently analysed web document or content is found again among the Top-10 search results which underlines and confirms this statement. Therefore, this tool can also be used to some extent to detect plagiarism in the WWW. Additionally, the user can easily modify the preselected query containing the most important keywords by clicking on them on the term lists or by using the query input field. Figure 5.5 shows a screenshot of DocAnalyser's result page with extracted search words (keywords and source topics) from the Wikipedia-article 'Dream' and Google's returned web results.

In its current Java-based implementation, DocAnalyser offers the following features:

- handling of different file formats such as PDF, DOC, DOCX, PPT, PPTX, HTML, XML and TXT
- language detection (German/English) and sentence splitting
- part-of-speech (POS) tagging (nouns, names, verbs, adjectives, adverbs, card.) and extraction of nouns and names
- phrase detection based on POS filters
- base form reduction
- high-quality keyword and source topic extraction based on the analysis of the selected contents' directed co-occurrence graphs

Another advantage of DocAnalyser's multi-threaded server is its ability to handle multiple user requests at a time. The graphical user interface of the result page as well as the bookmarklet to invoke DocAnalyser are both compatible with any modern web browser.

The service DocAnalyser offers goes beyond a simple search for similar documents as it presents a new way to search for related documents and documents with background information using source topics, too. This functionality can be regarded as a useful addition to services such as Google Scholar (http://scholar.google.com/), which offers users the possibility to search for similar scientific articles. It becomes especially useful when an in-depth research on a topic of interest has to be conducted. As the implemented extended HITS-based algorithm for keyword extraction employs semantic term relations for doing so, the quality of the obtained search words and the web search results gained using them is also high. Another noteworthy advantage of this keyword extraction algorithm is that it works on single texts as it does not rely on solutions based on term frequencies such as TF-IDF [8] and difference analysis [46, 136] that require preferably large reference corpora. As indicated before, in order to provide instant results, the (at most) four most important keywords/key phrases are automatically selected as search words which are sent to Google. The author chose four terms because, as discussed in Sect. 4.4.2, 4-term co-occurrences (when used as search words) provide the best trade-off between a precise search context description and a useful amount of search results. In this case, however, this 4-term co-occurrence (the query) is constructed for the complete web content provided, not only for a text fragment such as a sentence or paragraph.

Although this approach yields useful results, the user still has the option to interactively modify the preselected query. The extracted source topics provide another interesting use case, viz, the possibility to track topics to their roots (see Fig. 5.6) by iteratively using them as search words.

As the source topics of a text document represent its set of important influential aspects, the idea seems natural to (iteratively) use them as search words to find more documents in the WWW that in turn mainly deal with these topics. This approach is of benefit when an in-depth research on a topic of interest needs to be conducted. This way, it is possible to find related, not necessarily similar documents, too. The reason for this is the naturally occurring topic drift in the results induced by the semantic differences between authorities and hubs which become obvious when they are used as search words. Therefore, source topics of documents can be used as means to

Topic Tracking

Start | Google + HITS | Twitter + HITS | Google Standard | Twitter Standard

erneuerbare Energien | Search with HITS | Chin. Whispers | Voltage

Searchresults HITS + Chinese Whispers: erneuerbare Energien

Energiewende ? Wikipedia
http://de.wikipedia.org/wiki/Energiewende
Als **Energiewende** wird die Realisierung einer nachhaltigeren ..... Heizkraftwerken, durch die sich die **Energieverluste** im **Vergleich** zu im Kondensationsbetrieb ...

Erneuerbare Energien ? Wikipedia
http://de.wikipedia.org/wiki/Erneuerbare_Energien
Als **erneuerbare Energien** (fachsprachlich oftmals auch mit Majuskel: **Erneuerbare Energien**), regenerative Energien oder alternative Energien werden ...

Das Erneuerbare-Energien-Gesetz (EEG) | EEG aktuell
http://www.eeg-aktuell.de/das-eeg/
Das **Erneuerbare-Energien-Gesetz** (EEG) war und ist Motor und entscheidender Treiber für den Ausbau **erneuerbarer Energien** in Deutschland. Es trat im April ...

BMWi - Erneuerbare Energien auf einen Blick
http://www.bmwi.de/DE/Themen/Energie/Erneuerbare-Energien/erneuerbare-energien-auf-einen-blick.html
Windränder und Solaranlagen zum Thema **Erneuerbare Energien** ... Der Beitrag der **erneuerbaren Energien** zur Energieversorgung in **Deutschland** ist in den letzten ... Quelle: AG Energiebilanzen, Stand: Dezember 2013 (**PDF**: 158 KB).

Juwi: Bau von 120 neuen Windturbinen (ee-news.ch)
http://www.ee-news.ch/de/article/27873/juwi-bau-von-120-neuen-windturbinen
6. Jan. 2014 ... Insgesamt können die von der juwi-Gruppe seit der **Firmengründung** 1996 in Deutschland installierten Windenergie-Anlagen damit den ...

Windenergie ? Wikipedia
http://de.wikipedia.org/wiki/Windenergie
Windenergie bzw. **Windkraft** ist eine erneuerbare Energiequelle. Dabei wird die kinetische Energie von Wind, also bewegten Luftmassen der Atmosphäre, ...

Biomasse : Biomasse-Energie - So wird aus Biomasse ... - Die Welt
http://www.welt.de/wirtschaft/energie/specials/gas/article8796075/Biomasse-Energie-So-wird-aus-Biomasse-Energie.html
Über **Biomasse** lässt sich zum Beispiel elektrischer **Strom** erzeugen. Die Masse wird dazu zunächst verfeuert, mittels der **Wärmeenergie** wird dann in der Regel ...

Energiewende auf dem Prüfstand | ARTE Future
http://future.arte.tv/de/thema/energiewende
Das Jahrhundertprojekt **Energiewende** auf den Prüfstand: Warum geht der Ausbau .... Zum **Vergleich**: In Frankreich sind es aktuell gerade mal 128 GW (davon ...

Studie: 100 % Erneuerbare Energien für Strom und Wärme in ...
http://www.ise.fraunhofer.de/de/veroeffentlichungen/studien-und-positionspapiere/studie-100-erneuerbare-energien-fuer-strom-und-waerme-in-deutschland
In der **Studie** 100% **Erneuerbaren Energien** für Strom und Wärme in Deutschland untersuchen die Forscher des Fraunhofer ISE die Frage: "Wie könnte unsere ...

Fig. 5.6 DocAnalyser's Extension: Tracking the Topic 'Renewable Energies' (in German: erneuerbare Energien)

follow topics across several related documents. DocAnalyser implicitly supports this kind of search. Even so, DocAnalyser's solutions and approaches are just a first step towards a new way of searching the WWW, viz, the technically supported and librarian-like guidance of users to web documents and contents that actually fulfil their information needs. This approach will be elaborated on in subsequent chapters.

5.4.3 'PDSearch'—Finding Web Documents with Images

Additionally, most search engines today allow a search for pictures with amazing results. Either the content of pictures must be described by so-called metadata (which are again keywords describing contents and contexts) [138] or a search functionality for similar pictures (using their colour, contrast and other hard image values) is provided [120]. In fact, visual information plays an important role in humans life. The human brain is able to process complex picture information with a very high speed due to an extreme parallelisation [42]. A lot of details in a picture can be used to distinguish even similar contexts in a precise manner. Already in 1911, the expression 'Use a picture. It's worth a thousand words.' appears in a newspaper article by Arthur Brisbane discussing journalism and publicity [15]. The roots of that phrase are even older and have been expressed by earlier writers. For example, the Russian novelist Ivan Turgenev wrote (in Fathers and Sons in 1862), 'A picture shows me at a glance what it takes dozens of pages of a book to expound.'

Nevertheless, despite Google's own image search feature, there was no tool supporting a combined search for web documents and web images using pictures so far. Therefore, PDSearch (Picture-Document-Search) [117], an extension to DocAnalyser, has been designed and developed to specifically address this problem. Its aim is to support a search for web documents using the textual context (e.g. URL and description) of pictures in web pages that the user selects. In order to do so, PDSearch takes an initial query, returns matching results from Google's image search and allows users to select one or more of these images to refine the query and find more matching images or documents. The flow chart in Fig. 5.7 illustrates this algorithm.

The major goal of PDSearch was to assist the user in finding the correct meaning of short and ambiguous web queries by presenting matching and selectable web images. By clicking on one or more of these returned images (as shown on Fig. 5.8), the initial query can usually be disambiguated with a high precision. Furthermore, this kind of user feedback is both unobtrusive and relieves the user from having to reformulate a query manually when needed.

In the query refinement process, PDSearch analyses those web documents that contain the selected images (the images' contexts), extracts their four most relevant keywords, uses them as query terms and presents links to matching web documents using DocAnalyser's interface afterwards.

This approach is additionally extended by offering selectable keywords from the analysed web documents to be used as query terms. Generally, during the (first) search iterations using pictures as queries, a high recall of the search results is aimed at. When, however, the user is looking for matching web documents as seen in Fig. 5.9, this person can easily learn about new technical terms relevant for an in-depth research on a topic of interest. In this search step, the precision of the search results can be increased by the fine-grained selection of proper keywords as search words/query terms. Here again, the user does not need to enter these search words manually. A simple click on the presented keywords is sufficient to modify the current

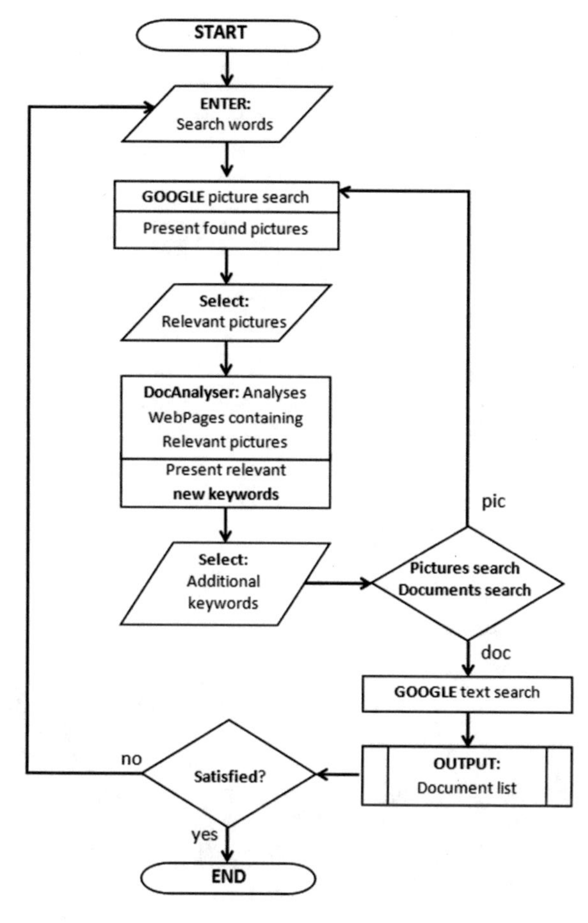

Fig. 5.7 The main algorithm of PDSearch

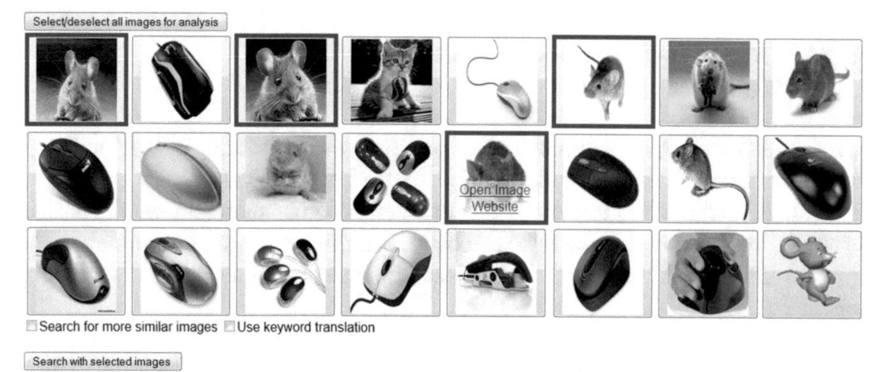

Fig. 5.8 PDSearch: Query refinement using images

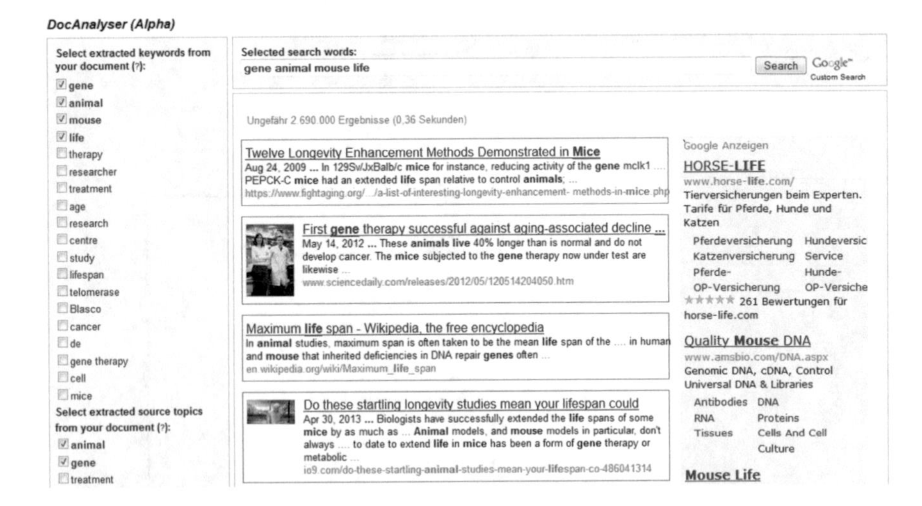

Fig. 5.9 DocAnalyser's result page

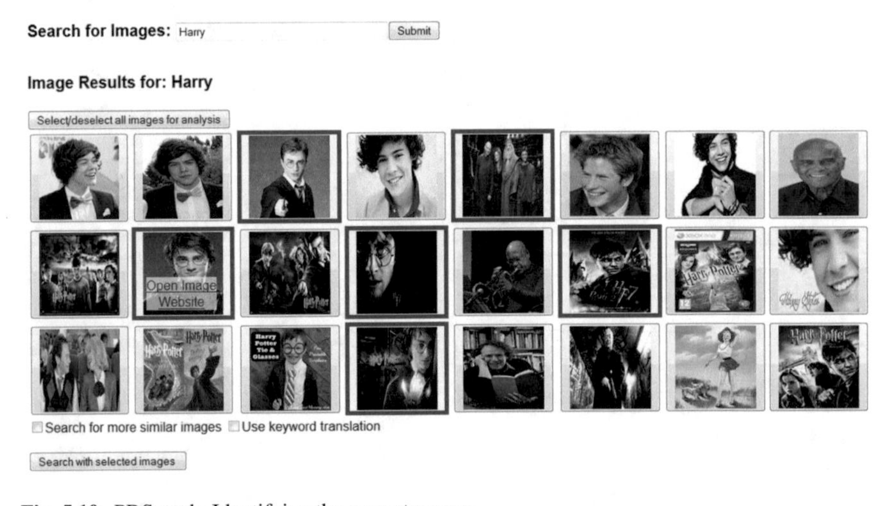

Fig. 5.10 PDSearch: Identifying the correct person

query, which is of great convenience during the search process. This approach is especially of benefit when the user needs to delve into a new topic.

While this solution is especially useful when it comes to determine the correct sense of short web queries, the same principle is helpful to identify the correct person given an incomplete name, e.g. when 'Harry' is used as a query, too. This can be seen in Fig. 5.10.

When the user is looking for more relevant web images instead of web documents, the respectively activated option would cause PDSearch to return matching web

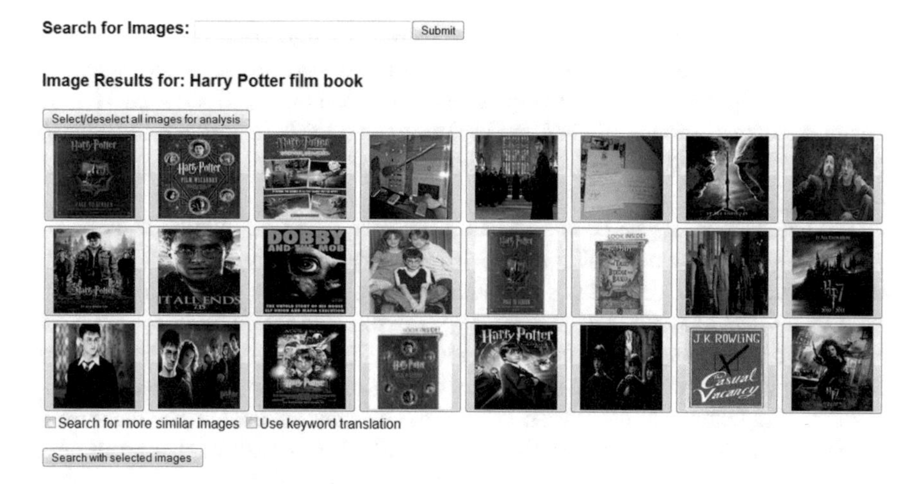

Fig. 5.11 PDSearch: Presenting more relevant images

images (see Fig. 5.11) as well, which—in turn—can be interactively selected again for another analysis.

In first user tests, the validity and flexibility of this search paradigm could be proven. However, as the text analysis process is one of the most important steps (it is responsible for the extraction of web documents' keywords and for the automatic construction of queries containing them) during the search sessions with PDSearch, it is sensible to think about options to increase the relevance of both kinds of search results (images and web documents). One way would be to realise a topically depending term weighting and clustering in conjunction with a named-entity recognition in order to improve the identification of relevant keywords and phrases. Another way would be to extend the already included cross-language search for English and German queries by adding an automatic translation from the initial and subsequent search queries into further languages. This way, relevant images and web documents in other languages can be found, too.

5.4.4 Android IR—The Mobile Librarian

So far, this chapter presented a number of locally working algorithms for context-based web search support dedicated to run on web servers and private computers. But what about mobile devices? Mobile technologies play an increasingly important role in everyday life. As an example, smartphones, tablets and even smartwatches are naturally used in both private and business sectors and can be regarded as centres of personal information processing as these portable devices concentrate diverse data and information streams and make them instantly and intuitively accessible by means of mobile applications (apps) running on them. While being able to also handle multimedia data with ease, these apps are mostly designed for the consumption and (to a lesser extent) the creation of textual contents. At the same time, the popularity of text-based apps is constantly rising: for instance, the number of active users of the instant messaging service WhatsApp [133] amounted to about 1 billion in February 2016. On 2016's New Year's Eve, 63 billion messages [88] have been sent using this service.

The analysis of large amounts of natural language text on mobile devices is—however—still uncommon, although their hardware is mostly powerful enough to carry out this task. Even today's mid-range smartphones are often equipped with more than 4 GB of RAM, fast multi-core processors and internal memory of 32 to 64 GB. Moreover, they are usually powered by a strong, rechargeable battery with 4000 mAh or more while at the same time their power-consumption is decreased by power-saving features of the operating system (OS) they run on. Despite these facts, current mobile devices along with their (pre-)installed apps mostly act as an intuitive terminal to request from or send data to remote services or servers, which usually carry out the (supposed to be) computationally expensive tasks. Economic interests play a significant role at this. For instance, companies can only gather insight from customer-related information when it is stored in their warehouses. However, it is sensible to offer local text processing solutions such as an integrated full-text search component that can run directly on mobile devices, especially smartphones. Several facts to underpin this assertion can be given:

- These devices concentrate information from different sources which can be enriched with context-related information provided by analysing data from the devices' sensors or adding relevant metadata from images on them to it.
- This unique composition of information as well as an in-depth analysis of it is likely of value to the user as she or he put it on or downloaded it to the mobile device in the first place.
- When the analysis can take place directly on the mobile device, the user's privacy is maintained as the information on her/his device does not have to be propagated to a possibly untrusted third party using an unsecured connection in order to use an analysis service.

When it comes to mobile information retrieval (mobile IR), in literature, a common point of view on this topic seems be that mobile devices only act as 'intelligent' search masks. As an example, Tsai et al. [119] use the term 'context awareness' to explain this view: the respective search context is augmented and enriched by the many features of the devices such as the integrated location functions. Practical examples for this approach are Apple's question answering system Siri, Microsoft's analogon Cortana, the music recognition app Shazam or one of the many location-based web services like Foursquare. A direct connection between the Android OS and mobile IR is presented in [112], which deals with natural language processing on Android devices, too. However, also in this case, the focus is put on the implementation of a library to invoke remote web services to analyse natural language text. The book 'Pocket Data Mining' [34] covers the highly interesting topic of distributed data mining on mobile devices and profoundly discusses the problems that come along with local (and mobile) data processing. In addition, it motivates these approaches and solutions by relevant real-world scenarios. However, as stated before, these approaches either neglect the fact that modern mobile devices are capable to handle large amount of data on their own or focus on the (nevertheless important) analysis of structured data only. Nonetheless, the autonomous processing of unstructured textual data on mobile devices is relevant, too. For this purpose, the Android OS is particularly suited as a target platform since

- the Android OS largely consists of open-source software and dominated the smartphone OS market with 85.0% share in Q1 of 2017 [131],
- Android apps are usually written in Java, making it possible to integrate existing Java-based libraries for natural language processing with minor adaptations and without having to completely reimplement them for the mobile application scenario and
- Android natively offers application programming interfaces (APIs) for parsing HTML files and with the library SQLite (https://www.sqlite.org/) an efficient solution for the persistent storage of mobile application data. A useful extension to SQLite, SQLite FTS (Full Text Search) (https://www.sqlite.org/fts3.html), enables the creation and query of inverted (word) indexes and can be used in Android as well, making it perfectly suited for the task at hand to create a functional full-text search solution for Android.

Therefore, the Android app 'Android IR',[2] which has been co-developed by the author, represents a first solution for effective, power-saving and completely integrated full-text search for Android devices. Android IR is able to extract, index and search text from local plain text documents as well as from PDF-and HTML-files stored on the device.

Besides the optimisation of these functionalities, power-saving features have been integrated as well which are helpful in conjunction with Android's own power management features. As it can be seen in Fig. 5.12, the user can activate energy options to only start the indexing service (which runs in the background even when the app's main activity is not visible) when the device is in sleep mode or connected to a charger. Furthermore, the service is only allowed to run when the current battery level is above a given threshold (default: 15%). In this context, the option to set the maximum number of pages of a PDF-file to be extracted (in menu 'text extraction') is helpful, too. Even more, in general settings, the user can specify the folders to be indexed as well as the maximum file size and the maximum processing time (for text extraction and indexing) in seconds allowed for one file. These options are, nevertheless, also intended to keep the local index small, whose maximum size can be adjusted in menu 'indexing'. Here, the user can also delete the entire index and select or deselect the option for stop word removal which is activated by default.

As the overall performance of Android IR has been positively evaluated (e.g. the device Nexus 9 needed 26 min to index 1.2 GB of HTML-and PDF-files and

---

[2]Interested readers may download 'Android IR' (16.8 MB; installation of apps from unknown sources must be allowed in security settings) from: http://www.docanalyser.de/androidir.apk.

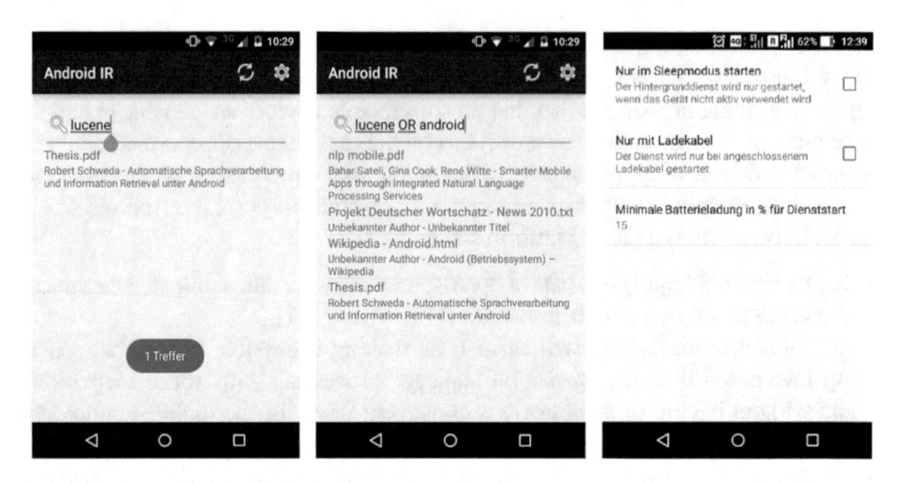

Fig. 5.12 Screenshots of Android IR: Search result lists and energy options

its battery level decreased by only 3% in doing so), current works are dedicated to extend Android IR to turn it into a powerful 'mobile librarian or secretary' which can autonomously combine, analyse and correlate textual data from various sources in order to act as a helpful personal information manager (PIM), e.g. extract appointments and remind the user of them. For doing so, it is necessary to be able to correctly identify named entities such as mentioned persons, organisations, locations as well as date-and time-related data. The already mentioned library Apache OpenNLP would be good choice for this task. Its applicability as part of a mobile app is currently being investigated. Furthermore, the usage of text mining methods to improve the app's search functionalities would be beneficial. As the demand for autonomous data analysis solutions on mobile devices will definitely grow in the future, the development of such methods is necessary, not only in the context of IR solutions. From the technological point of view, it is clearly possible (and sensible) to realise efficient and effective mobile search solutions that are backed by both flexible natural language and text mining tools as well as local database systems. In this regard, the app Android IR is just a first, nevertheless important, step towards a holistic search solution for mobile devices.

5.5 Summary

This chapter introduced locally working algorithms, tools and services to improve the search for documents in the web as well as on mobile devices. These solutions have been thoroughly tested and evaluated regarding their effectiveness. Although they have been realised as stand-alone applications, their individual components provide well-defined interfaces and can be easily integrated into the 'WebEngine' to provide their features and functions in this decentralised web search engine as well. One of the key aspects in all solutions was to provide means to guide users to relevant documents (or images). This line of thought shall be picked up again by the new concepts to realise the 'Librarian of the Web' which will be introduced in the next three chapters.

Chapter 6
Text-Representing Centroid Terms

6.1 Fundamentals

For the subsequent text analysis approaches presented in this book, the usage of co-occurrences and co-occurrence graphs (see Fig. 6.1) has been selected. They are an ideal means to obtain more detailed information about text documents than simple term frequency vectors or similar approaches could ever offer.

Co-occurrence graphs can indicate the strength of semantic connections of terms in analysed text corpora and can thus be regarded a proper knowledge base for further text analysis steps.

As mentioned before, two words w_i and w_j are called *co-occurrents*, if they appear together in close proximity in a document D. A *co-occurrence graph* $G = (W, E)$ may be obtained, if all words W of a document or set of documents are used to build its set of nodes which are then connected by an edge $(w_a, w_b) \in E$ if $w_a \in W$ and $w_b \in W$ are co-occurrents. A weight function $g((w_a, w_b))$ indicates, how significant the respective co-occurrence is in a document. If the significance value is greater than a pre-set threshold, the co-occurrence can be regarded as significant and a semantic relation between the words involved can often be derived from it.

As mentioned before, significance measures that determine the value of $g((w_a, w_b))$ for given words w_a and w_b will generate undirected co-occurrence graphs. Directed co-occurrence graphs (also called association graphs) can be obtained by applying the well-known method of association rule learning from the field of data mining. In the following elaborations, undirected co-occurrence graphs will be focussed.

A co-occurrence graph—similarly to the knowledge in the human brain—may be built step by step over a long time taking one document after another into consideration. From the literature [46] and own experiments (see Fig. 6.2) it is known that the outdegrees of nodes in co-occurrence graphs follow a power-law distribution and the whole graph exhibits small-world properties with a high clustering coefficient as well as a short average path length between any two nodes. This way, a co-occurence graph's structure also reflects the organisation of human lexical knowledge.

M. Kubek, *Concepts and Methods for a Librarian of the Web*,
Studies in Big Data 62, https://doi.org/10.1007/978-3-030-23136-1_6

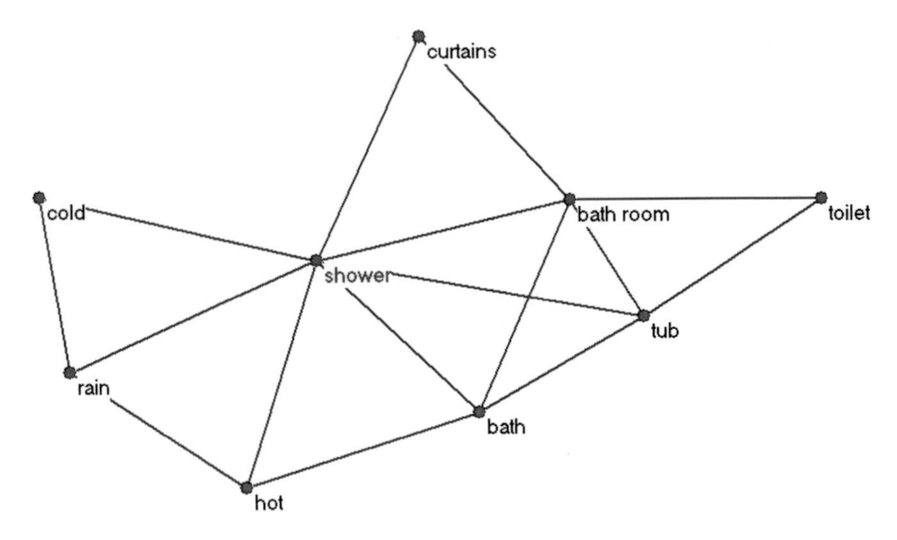

Fig. 6.1 A co-occurrence graph for the word 'shower'

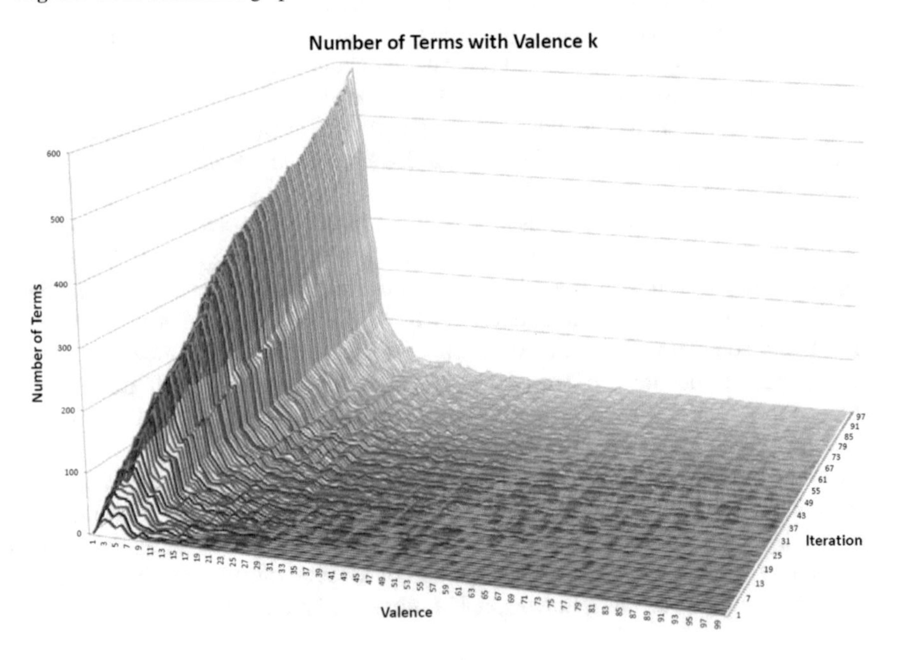

Fig. 6.2 Distribution of outdegrees in a co-occurrence graph over time

The use of the immediate neighbourhood of nodes in a co-occurrence graph has been widely considered in literature, e.g. to cluster terms [11], to determine the global context (vector) of terms in order to evaluate their similarity [46] and to detect temporal changes in this vector [45] to make a term's change of meaning over time visible or to derive paradigmatic relations between two terms [12]. This understanding shall be extended now, since the presentation of thoughts e.g. during human speech production usually relies on building chains of collocations and co-occurrences. In doing so, multiple words are connected over distances ≥ 2 in the co-occurrence graph. Thus, indirect neighbourhoods of terms in co-occurrence graphs (nodes that can be reached only using two or more edges from a node of interest) and the respective paths with a length ≥ 2 should be considered as well as indirectly reachable nodes may still be of topical relevance, especially when the co-occurence graph is large. The benefit of using such nodes/terms in co-occurrence graphs has already been shown for the expansion of web search queries using a spreading activation technique applied on local and user-defined corpora [71]. The precision of web search results can be noticably improved when taking those terms into account, too.

The field of application of indirect term neighbourhoods in co-occurrence graphs shall be extended in the next section by introducing an approach to determine centroid terms of text documents that can act as their representatives in further text processing tasks. These centroid terms can be regarded as the texts' topical centres of interest (a notion normally used to describe the part of a picture that attracts the eye and mind) that the author's thoughts revolve around.

6.2 Finding Centroid Terms

In physics, complex bodies consisting of several single mass points are usually represented and considered by their so-called centre of mass, as seen in Fig. 6.3. The distribution of mass is balanced around this centre and the average of the weighted coordinates of the distributed mass defines its coordinates and therefore its position.

For discrete systems, i.e. systems consisting of n single mass points $m_1, m_2, \ldots, m_i$ in a $3D$-space at positions $\mathbf{r}_1, \mathbf{r}_2, \ldots, \mathbf{r}_i$, the centre of mass $\mathbf{r}_s$ can be found by

$$\mathbf{r}_s = \frac{1}{M} \sum_{i=1}^{n} m_i \mathbf{r}_i, \tag{6.1}$$

whereby

$$M = \sum_{i=1}^{n} m_i. \tag{6.2}$$

Usually, this model simplifies calculations with complex bodies in mechanics by representing the whole system by a single mass at the position of the centre of mass. Exactly the same problem exists in automatic text processing: a whole text shall be

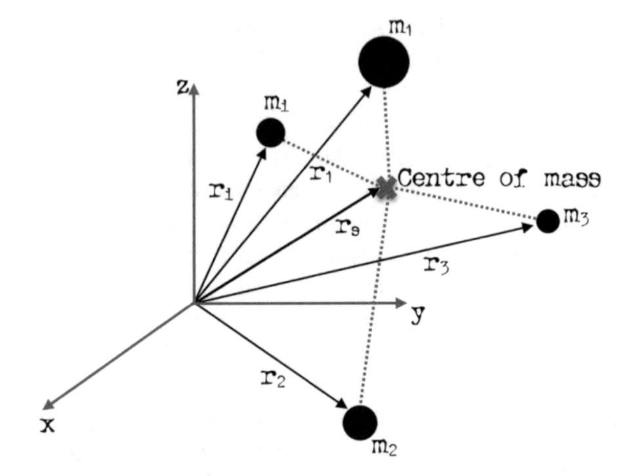

Fig. 6.3 The physical centre of mass

represented or classified by one or a few single, descriptive terms which must be found.

To adapt the situation for this application field, first of all, a *distance d* shall be introduced in a co-occurrence graph G. From the literature it is known that two words are semantically close, if $g((w_a, w_b))$ is high, i.e. they often appear together in a sentence or in another predefined window of n words. Consequently, a distance $d(w_a, w_b)$ of two words in G can be defined by

$$d(w_a, w_b) = \frac{1}{g((w_a, w_b))}, \tag{6.3}$$

if w_a and w_b are co-occurrents. In all other cases (assuming that the co-occurrence graph is connected[1]) there is a shortest path $p = (w_1, w_2), (w_2, w_3), \ldots, (w_k, w_k + 1)$ with $w_1 = w_a$, $w_{k+1} = w_b$ and $w_i, w_{i+1} \in E$ for all $i = 1(1)k$ such that

$$d(w_a, w_b) = \sum_{i=1}^{k} d((w_i, w_{i+1})) = MIN, \tag{6.4}$$

whereby in case of a partially connected co-occurrence graph $d(w_a, w_b) = \infty$ must be set. Note, that differing from the physical model, there is a distance between any two words but no direction vector, since there is no embedding of the co-occurrence graph in the 2- or 3-dimensional space. Consequently, the impact of a word depends only on its scalar distance.

[1]This can be achieved by adding a sufficiently high number of documents to it during its building process.

In continuation of the previous idea, the distance between a given term t and a document D containing N words $w_1, w_2, \ldots, w_N \in D$ that are reachable from t in G can be defined by

$$d(D,t) = \frac{\sum_{i=1}^{N} d(w_i,t)}{N}, \tag{6.5}$$

i.e. the average sum of the lengths of the shortest paths between t and all words $w_i \in D$ that can be reached from it. Note that—differing from many methods found in the literature—it is not assumed that $t \in D$ holds!

The centroid term $\chi(D)$ is defined to be the term with

$$d(D, \chi(D)) = MINIMAL. \tag{6.6}$$

Also, it might happen in some cases that the minimal distance is not uniquely defined. Consequently, a text may have more than one centroid term (as long as no other methods decide which one is to use).

In order to define the centroid-based distance ζ between any two documents D_1 and D_2, let $\chi(D_1)$ be the centre term or *centroid term of* D_1 with $d(D_1, \chi(D_1)) = MIN$. If at the same time $\chi(D_2)$ is the centroid term of D_2,

$$\zeta(D_1, D_2) = d(\chi(D_1), \chi(D_2)) \tag{6.7}$$

can be understood as the semantic distance ζ of the two documents D_1 and D_2. In order to obtain a similarity value instead,

$$\zeta_{sim}(D_1, D_2) = \frac{1}{1 + \zeta(D_1, D_2)} \tag{6.8}$$

can be applied.

It is another important property of the described distance calculation that documents regardless of their length as well as short phrases can be assigned a centroid term by one and the same method in a unique manner. The presented approach relies on the preferably large co-occurrence graph G as its reference. It may be constructed from any text corpus in any language available or directly from the sets of documents whose semantic distance shall be determined. The usage of external resources such as lexical databases or reference corpora is common in text mining: as an example, the so-called difference analysis [46, 136] which measures the deviation of word frequencies in single texts from their frequencies in general usage (a large topically well-balanced reference corpus is needed for this purpose) is an example for it. The larger the deviation is, the more likely it is that a term or keyword of a single text has been found.

Table 6.1 Centroids of 30 Wikipedia articles

Title of Wikipedia article	Centroid term
Art competitions at the Olympic games	Sculpture
Tay-Sachs disease	Mutation
Pythagoras	Pythagoras
Canberra	Canberra
Eye (cyclone)	Storm
Blade Runner	Ridley Scott
CPU cache	Cache miss
Rembrandt	Louvre
Common Unix printing system	Filter
Psychology	Psychology
Religion	Religion
Universe	Shape
Mass media	Database
Rio de Janeiro	Sport
Stroke	Blood
Mark Twain	Tale
Ludwig van Beethoven	Violin
Oxyrhynchus	Papyrus
Fermi paradox	Civilization
Milk	Dairy
Corinthian war	Sparta
Health	Fitness
Tourette syndrome	Tic
Agriculture	Crop
Finland	Tourism
Malaria	Disease
Fiberglass	Fiber
Continent	Continent
United States Congress	Senate
Turquoise	Turquoise

It is now sensible to get a first impression of the centroid terms' quality with respect to whether they are actual useful representatives of text documents. Table 6.1 therefore presents the centroid terms of 30 English Wikipedia articles. The corpus used to create the reference co-occurrence graph G consisted of 100 randomly selected articles (including the mentioned 30 ones) from an offline English Wikipedia corpus from http://www.kiwix.org. In Sect. 6.5, the experimental setup and datasets used along with an extension of this experiment will be explained in more detail.

It can be seen that almost all centroids properly represent their respective articles.

6.3 Spreading Activation: A Fast Calculation Method for Text Centroids

The direct application of the previous definition within an algorithm would require to check all nodes of the fully connected (sub-)component of the co-occurrence graph, whether they fulfil the above defined minimum property. This results in an average complexity of $\mathbf{O}(|W|^3)$. Since co-occurrence graphs may contain even more than 200,000 nodes (including nouns, names, composites etc.), significant calculation times in the range of several minutes (see Table 6.2 in Sect. 6.3.3) may appear even on powerful machines.

Therefore, a slight adaptation may be indicated which additionally avoids computationally expensive division operations. In the following, $\chi(D)$ denotes the centroid term of a document D, the term which minimises the longest distance to any of the words in the document, i.e. $d'(D, \chi(D)) = MINIMAL$ for all $t \in W(D)$ and

$$d'(D, t) = MAX(d(w_i, t)|i = 1 \dots N).$$

In numerous experiments, it was found that the deviations caused by this adaptation are not too big. With these changes, a much faster, locally working method to determine the centroid of any document can be presented that only affects a certain local area of the co-occurrence graph G which is relevant for the query or text documents to be analysed and in doing so avoids the check of all nodes in the graph. The following method utilises the spreading activation technique [98] to address this problem.

In order to understand its working principle, the physical correspondence of the text centroids, i.e. the centre of mass must be considered again. In a physical body, the centre of mass is usually expected to be inside a convex hull line of the (convex or concave) body, in case of a homogeneous one and is situated more or less in its middle (Fig. 6.4[2]).

If a set of uniform, discrete mass points in a 2-dimensional, Euclidean plane with an underlying rectangular grid is considered, one would try to fix the centre of mass in the intersection of concentric cycles of the same radius around those points (Fig. 6.5).

Things may look more complex in a usual co-occurrence graph, since it can normally not be embedded in a 2- or 3-dimensional space, which humans can easily imagine. However, similar ideas of a neighbourhood allocation have already been

Table 6.2 Processing times of the original algorithm

Number of query terms	2	3	4	5	6
Processing time in [s]	185	252	317	405	626

[2]Modified from https://commons.wikimedia.org/wiki/File:Bird_toy_showing_center_of_gravity.jpg, original author: APN MJM, Creative Commons licence: CC BY-SA 3.0.

Fig. 6.4 A convex hull curve of a bird-toy and its centre of mass

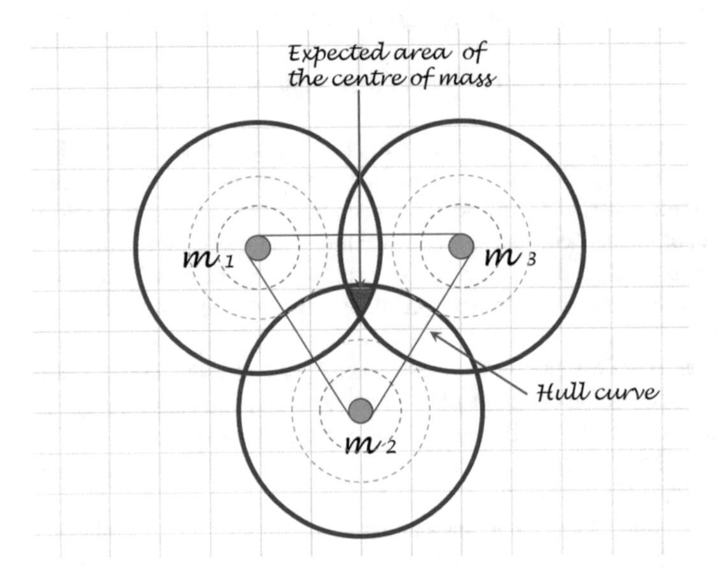

Fig. 6.5 Locating the centre of mass in the 2-dimensional plane

used to provide a graph clustering method [121]. The already mentioned Chinese Whispers algorithm [11] is another interesting and related solution for efficient graph clustering that relies on a label propagation technique. The following algorithm adapts the idea described above and can be applied on large co-occurrence graphs quite fast.

6.3.1 Algorithm

Usually, the co-occurrence graph can be completely kept on one machine. Therefore, the calculation of text centroids can be carried out in a local, serial manner. If only shortest paths are considered between any two nodes in the co-occurrence graph, a metric system is built.

In the following considerations, a query set Q of s words $Q = \{w_1, w_2, \ldots, w_s\}$ shall be considered. Q is called a **query set**, if it contains (usually after a respective preprocessing) only words $w_1, w_2, \ldots, w_s$, which are nodes within a single, connected component of the co-occurrence graph $G = (W, E)$ denoted by $G' = (W', E')$.

For computation purposes, a vector $\bar{v}(w') = [v_1, v_2, \ldots, v_s]$ is assigned to each $w' \in W'$ with the components being initialised to 0. With this preparations, the following, **spreading activation algorithm** is executed.

1. Determine (or estimate) the maximum of the shortest distances d_{max} between any pair (w_i, w_j) with $w_i, w_j \in Q$, i.e. let

$$d_{max} = sup(d(w_i, w_j)|(w_i, w_j) \in Q \times Q)$$

2. Choose a radius $r = \frac{d_{max}}{2} + \Delta$, where Δ is a small constant of about $0.1 \cdot d_{max}$ ensuring that an overlapping area will exist.
3. Apply (or continue) a breadth-first-search algorithm from every $w_i \in Q$ and activate (i.e. label) each reached, recent node w' for every w_i by

$$\bar{v}(w')[v_i] = d(w_i, w') \leftrightarrow d(w_i, w') \leq r.$$

   Stop the activation, if no more neighbouring nodes with $d(w_i, w') \leq r$ can be found.
4. Consider all nodes $w' \in W'$ with

$$\forall i, \quad i = 1 \ldots s : \bar{v}(w')[v_i] \neq 0$$

   and choose among them the node with the **minimal**

$$MAX(\bar{v}(w')[v_i])$$

   to be the centroid $\chi(\{w_1, w_2, \ldots, w_s)\}$.
5. If no centroids found, set $r := r + \Delta$ and GoTo 3, otherwise *STOP*.

The greatest benefit of the described method is that it generally avoids the 'visit' of all nodes of the co-occurrence graph as it solely affects local areas around the query terms $w_1, w_2, \ldots, w_s \in Q$.

Also, as the employed spreading activation technique can be independently executed for every single initially activated node/term (e.g. from a query), the algorithm's core steps can be performed in parallel, e.g. in separate threads. This makes an effective utilisation of potentially available multiple CPU cores possible.

6.3.2 Diversity and Speciality

The supremum

$$\mu = sup(d(w_i, w_j)|(w_i, w_j) \in (Q \times Q)), \tag{6.9}$$

which was mentioned in the algorithm, is called the **diversity** of the query set Q. The smaller the diversity is, the more a query targets a designated, narrow topic area, while high values of the diversity mark a more general, common request.

Here, it is assumed that the distance $d(w_i, w_j)$ of any two words w_i, w_j of a query set Q in the respective reference co-occurrence graph G can be calculated. It is also assumed that all words $w_i, w_j \in Q$ are nodes of that co-occurrence graph.

The term

$$\sigma = \frac{1}{1 + sup(d(\chi_G(Q), w_i)|w_i \in Q)} \tag{6.10}$$

obtained by calculating the distance of the centroid $\chi_G(Q)$ to the query set Q with all contained words w_i is called its **speciality** σ.

Diversity and speciality are related to each other, whereby the former mainly expresses the most extreme (semantic) differences in a text/query and the latter focusses on the largest distance of the centroid term of Q to any contained word w_i. These two measures are especially helpful to determine the quality of a query in interactive search sessions. This field of application for them will be discussed in detail in Chap. 8.

6.3.3 Experimental Evaluation

In this section, the performance of the presented algorithm will be evaluated in a number of experiments. All measurements have been performed on a Lenovo Thinkpad business-class laptop equipped with an Intel Core i5-6200U CPU and 8 GB of RAM to show that the algorithm can even be successfully applied on non-server hardware. The four datasets[3] used to construct the co-occurrence graphs consist of either 100, 200, 500 or 1000 topically classified (topical tags assigned by their authors) online news articles from the German newspaper 'Süddeutsche Zeitung'.

[3]Interested readers may download these datasets (4.1 MB) from: http://www.docanalyser.de/sa-corpora.zip.

In order to build the (undirected) co-occurrence graphs, linguistic preprocessing has been applied on these documents whereby sentences have been extracted, stop words have been removed and only nouns (in their base form), proper nouns and names have been considered. Based on these preparatory works, co-occurrences on sentence level have been extracted. Their significance values have been determined using the Dice coefficient [28]. These values and the extracted terms are persistently saved in an embedded Neo4j (https://neo4j.com) graph database using its property-value store provided for all nodes (represent the terms) and relationships (represent the co-occurrences and their significances).

Experiment 1: Average Processing Time

In the first sets of experiments for this algorithm, the goal is to show that for automatically generated queries of five different sizes (queries consisting of two to six terms) in six different ranges of diversity the average processing time to find their centroid terms is low when the spreading activation method is applied.

The queries and the used co-occurrence graph G have been generated using the dataset 'Corpus-100' from which 4331 terms have been extracted. In order to determine the average processing time for each query size and for the six ranges of diversity, 20 different queries have been generated for these two parameters. Therefore, altogether 600 queries have been created. Figure 6.6 shows that even for an increased number of query terms and diversity values the average processing time stays low and increases only slightly. As the average processing time stays under half a second for all cases, the algorithm is clearly suited for application in interactive search systems.

In order to demonstrate the great improvement in processing time of this algorithm, the original algorithm (as the direct application of the centroid definition given in Sect. 6.2) has been run on five queries consisting of two to six terms in the diversity range of [10–15) as a comparison while using the dataset 'Corpus-100' to construct the co-occurrence graph G as well. Table 6.2 presents the absolute processing times in

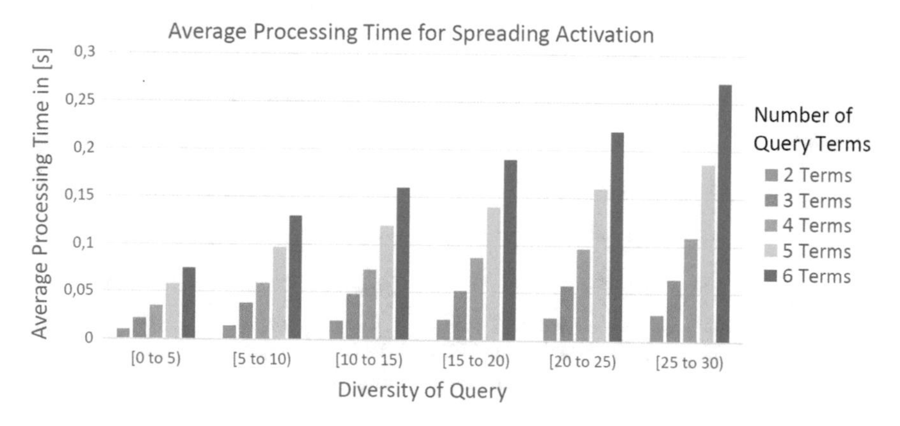

Fig. 6.6 Average processing time for spreading activation

[s] needed by the original algorithm to determine the centroid terms for those queries. Due to these high and unacceptable values, the original algorithm—in contrast to the one presented in the previous section—cannot be applied in interactive search systems that must stay responsive at all times.

Experiment 2: Node Activation

The second set of experiments examines the average number of nodes activated when the presented algorithm is run starting from a particular centroid term of a query while restricting the maximum distance (this value is not included) from this starting node. As an example, for a maximum distance of 5, all activated nodes by the algorithm must have a smaller distance than 5 from the centroid. In order to be able to determine the average number of activated nodes, the algorithm has been started for 10 centroid terms under this restriction. The maximum distance has been varied (increased) from 1 to 25. For this experiment, the dataset 'Corpus-100' has been used again.

As Fig. 6.7 shows, the average number of activated nodes visibly and constantly rises starting from the maximum distance of 7 (40 activated nodes). At the maximum distance of 25, in average, 3780 nodes have been activated. The result also shows that for queries with a low diversity (and a therefore likely high topical homogeneity) the number of activated nodes will stay low as well. In this example, for a low maximum distance of 10, the average number of activated nodes is only 88 (2% of all nodes in the used co-occurrence graph G). Therefore and as wished-for, the activation stays local, especially for low diversity queries.

Experiment 3: Growing Co-occurrence Graph

As document collections usually grow, the last sets of experiments investigate the influence of a growing co-occurrence graph on the processing time of the introduced

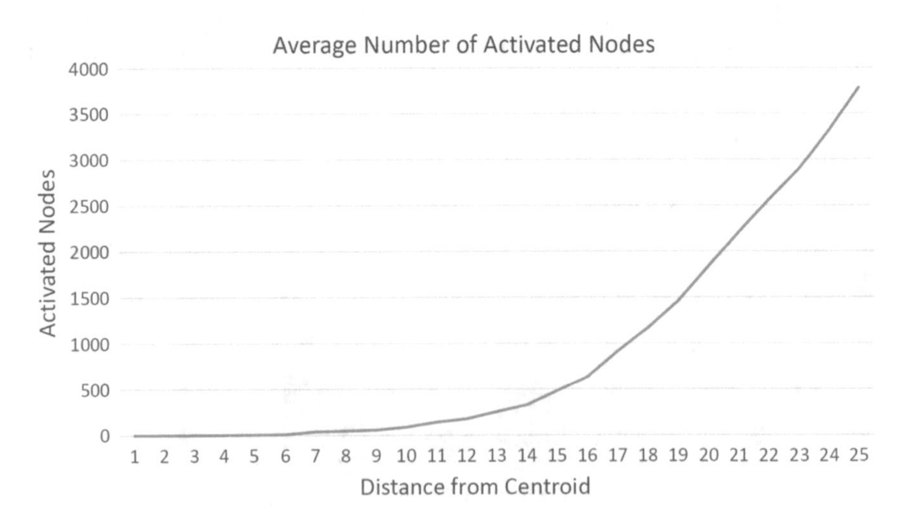

Fig. 6.7 Average number of activated nodes

algorithm. For this purpose, the datasets 'Corpus-100', 'Corpus-200', 'Corpus-500' and 'Corpus-1000' respectively consisting of 100, 200, 500 and 1000 news articles have been used to construct co-occurrence graphs of increasing sizes with 4331, 8481, 18022 and 30048 terms/nodes. Also, as the document collection should be growing, it is noteworthy to point out that corpora of smaller sizes are included in the corpora of larger sizes. For instance, the articles in dataset 'Corpus-200' are included in both 'Corpus-500' and 'Corpus-1000', too. In order to conduct the experiments, four queries have been chosen: one query with two terms and a low diversity in the range of [5–10), one query with two terms and a high diversity in the range of [20–25), one query with six terms and a low diversity in the range of [5–10) and one query with six terms and a high diversity in the range of [20–25). For each of these queries and each corpus, the absolute processing time in [s] (in contrast to the previous experiments that applied averaging) to determine the respective centroid term has been measured.

The curves in Fig. 6.8 show an almost linear rise in processing time for all four queries and the four growing co-occurrence graphs. Besides the size of the co-occurrence graph G used, the query size is of major influence on the processing time. While the query's diversity plays a rather secondary role at this, it can clearly be seen that—even initially—the processing times for the queries with a high diversity are higher than for their equal-sized counterparts with a low diversity. However, even in these experiments and especially for the query with six terms and high diversity and the largest co-occurrence graph of 30048 nodes, the processing time stayed low with 0.41 s. While these experiments showed that the processing time will understandably increase when the underlying co-occurrence graph is growing, its rise is still acceptable, especially when it comes to handle queries in a (graph-based) search system. The reason for this is again the algorithm's local working principle. Node activation will occur around the requested query terms only while leaving most of the nodes in the graph inactivated.

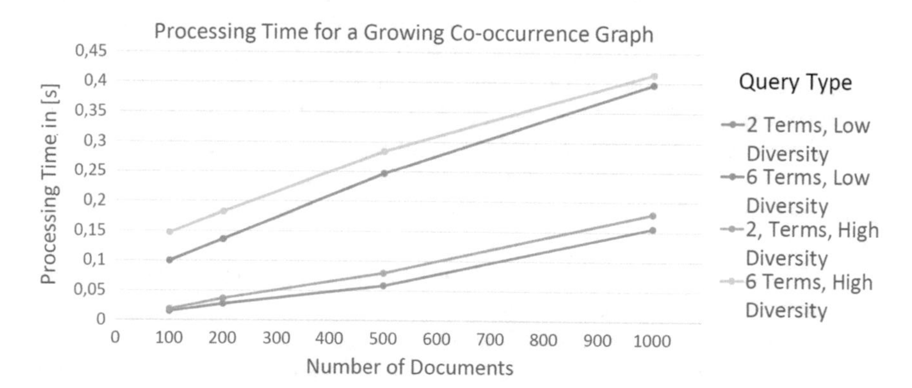

Fig. 6.8 Influence of a growing co-occurrence graph

6.4 Properties of Co-occurrence Graphs and Centroid Terms

This section presents important properties of co-occurrence graphs and centroid terms. They are of relevance for the modelling of the human learning process outlined at the beginning of Chap. 4 as well as for the motivation and justification of the considerations in the following chapters. For instance, it is necessary that centroid terms do not change frequently or make expected 'jumps' (after subsequent calculations) even when the underlying co-occurrence graph is growing or the strength of its term relationships varies due to incoming documents. The stability of both co-occurrence graphs and centroid terms is necessary to make continuously correct classification decisions. These properties will be confirmed by simulative experiments.

For all of the exemplary experiments discussed in this section, linguistic preprocessing has been applied on the documents to be analysed whereby stop words have been removed and only nouns (in their base form), proper nouns and names have been extracted. In order to build the undirected co-occurrence graph G (as the reference for the centroid distance measure), co-occurrences on sentence level have been extracted. Their significance values have been determined using the Dice coefficient [28]. The particularly used sets of documents to create G and to calculate the centroid terms will be described in the respective subsections.[4]

6.4.1 *Stability of the Co-occurrence Graph*

The first experiment shall confirm the fast convergence and stability of the co-occurrence graph which is a prerequisite for its use as a dynamic knowledge base of the individual or local computing node. For this purpose, a co-occurrence graph is constructed from a text corpus in an iterative manner by successively adding co-occurrences from one document after another and finally removing all non-significant co-occurrence relations. In this experiment, the topically well-balanced corpus from dataset 'CGGrowth' containing 100 randomly chosen online news articles from the German newspaper 'Süddeutsche Zeitung' from the months September, October and November of 2015 and covers 19 topics and the topically focussed corpus from the same dataset containing 100 articles on the European migrant crisis (a hotly discussed topic in late 2015) from the same newspaper and the same time period have been used.

The first mentionable result is that during this learning process, the number of nodes and edges added to the co-occurrence graph G for each new incoming document is converging. As especially high-weight edges representing significant co-occurrences are of interest for the centroid determination, Fig. 6.9 shows that the

[4]Interested readers may download these datasets (1.03 MB) from: http://www.docanalyser.de/cd-prop-corpora.zip.

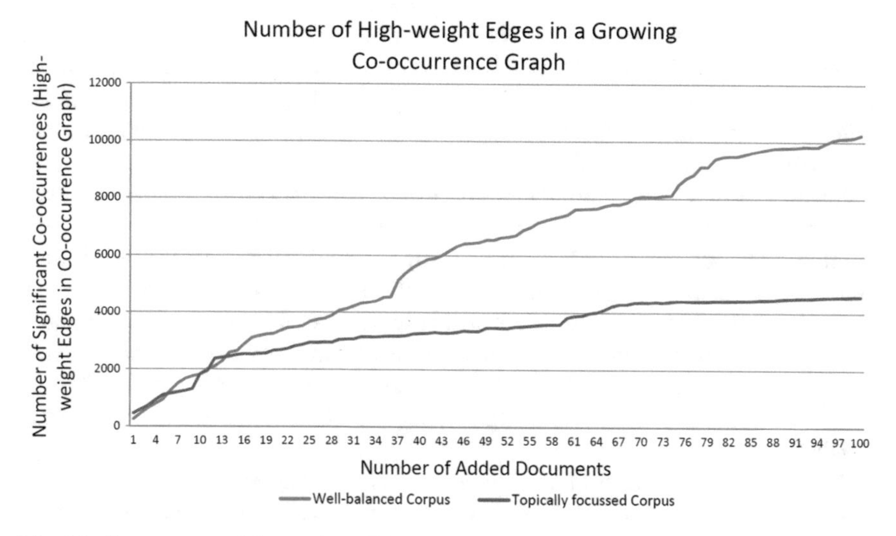

Fig. 6.9 Convergence of the number of high-weight edges in a growing co-occurrence graph

number of these edges converges when the topically well-balanced corpus is used to construct G. The effect is even stronger for the topically focussed corpus as the terminology used in it does not vary greatly.

Additionally, the probability that new words (nodes) are included is drastically decreasing; also, the nodes' ranks (according to their outdegrees) only seldom change. The node with rank 1 has the highest connectivity. Figure 6.10 shows that

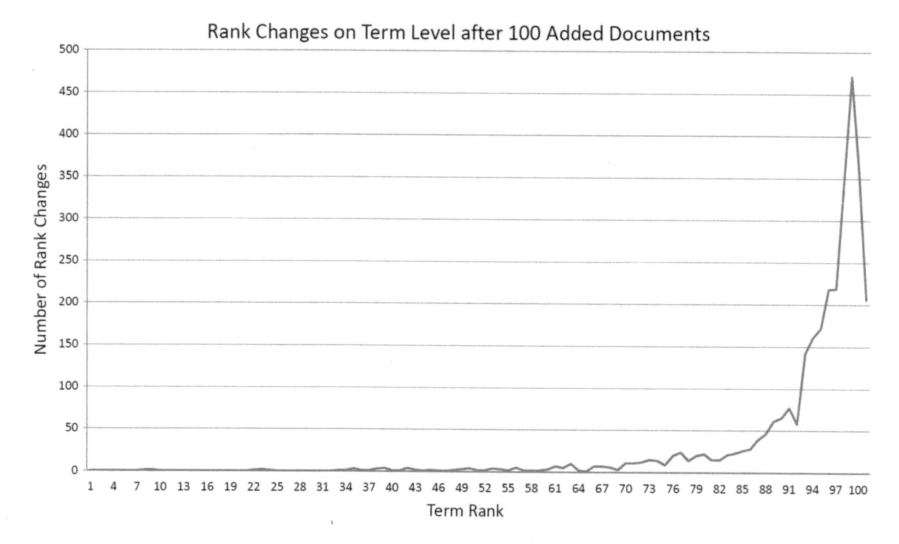

Fig. 6.10 Stability of term ranks in a grown co-occurrence graph

most of the rank changes occur at the nodes with higher ranks (in this case, rank changes have been determined after 100 documents have been added to the collection) only. These are usually nodes with low connectivity and often have been added to the co-occurrence graph just recently. For this experiment, the topically focussed corpus from dataset 'CGGrowth' has been used again.

As mentioned before, centroid terms normally should not change frequently, even when the co-occurrence graph is growing. To confirm this hypothesis, for this experiment, a reference document's centroid term has been calculated after each time step. Also, in each time step, a single document has been added to the co-occurrence graph. The corpus from dataset 'CentroidMovement' used for this experiment contains 100 newspaper articles from 'Süddeutsche Zeitung' which cover three topical categories 'car' (34 articles), 'finance' (33 articles) and 'sports' (33 articles). The reference document used was 'Schmutzige Tricks' (an article on the 2015 car emissions scandal).

As it can be seen in Fig. 6.11, the 'movement' of the reference document's centroid term (calculated after each time step) stabilises quickly. However, the convergence time depends on the order in which documents are added to the co-occurrence graph. Here, the topical orientation and similarity between these documents play an important role. If similar documents to the reference document are added first (blue curve) and other, topically dissimilar documents afterwards, then the centroid term changes rarely (almost never) during their addition. If, however, the documents are added randomly (orange curve), then the convergence time increases. The reason for this observation is that topically similar documents (which mostly influence the centroid term's position) can be added at any time. Thus, the probability that the centroid term changes at any time is increased, too.

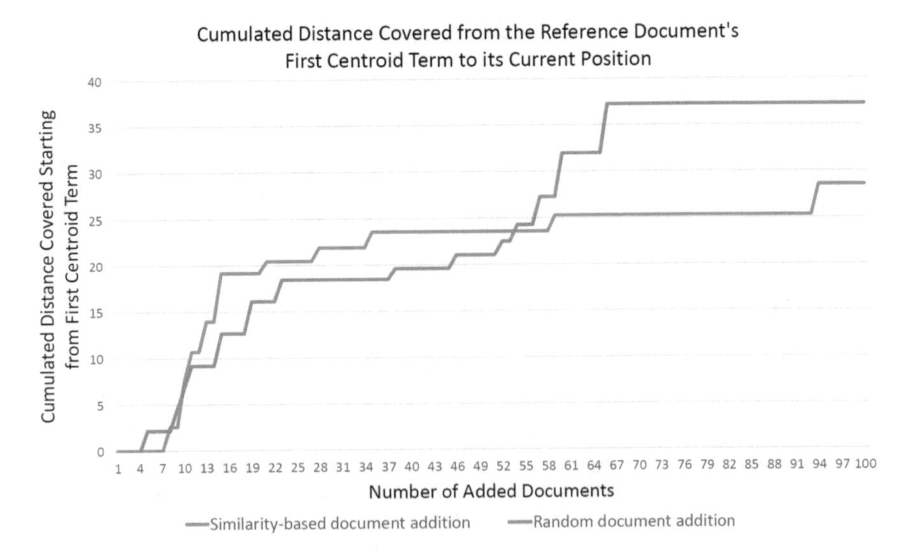

Fig. 6.11 Movement of a reference document's centroid term in a growing co-occurrence graph

Therefore, it is not surprising that mainly in the first 34 time steps during the similarity-based document addition (in which the 34 car-related documents are added) the centroid term's position of this article is changed. Even so, in both cases, the centroid term's jumping distance between two consecutive 'positions' in the co-occurrence graph is low. However, it is still possible that the centroid term (usually just slightly) changes when the co-occurrence graph significantly grows as this process changes the distances between all nodes as well.

Additionally, as a co-occurrence graph reflects the organisation of human lexical knowledge, these observations might give an interesting insight into the learning processes of the human brain and might bear an explanation why for many people learning new skills such as the acquisition of a second language is hard. As new concepts/terms/words are automatically matched with the well-known and stable semantic contexts of those previously learned (especially with those in their mother tongue) which these additions usually do not (yet) have, their own 'semantic position' as well as their semantic contexts are not fixed yet (their ranks change often). It will take some time to find the right 'position' and the right 'neighbours' to connect to in order to be proficient in a certain field e.g. a foreign language. Therefore, a constant practise (in accordance to the continuous presentation of similar documents) is needed for this purpose, especially when some knowledge or a conception of related terms already exists in order to avoid confusion. Further research and investigations are needed to evaluate these findings properly and probably derive a general-purpose model which suitably reflects the learning processes in the human brain.

6.4.2 Uniqueness of Centroids

Theoretically, it might happen in a given co-occurrence graph that one or more terms have the same, minimal average distance to all terms of a text document. This would mean that the centroid term is not uniquely defined and more than one term could represent the document. In particular, this complies with reality as some documents, especially interdisciplinary ones, may not by clearly assigned to the one or another category. However, the subsequently explained practical experiences justify that this case in fact might only appear extremely rarely.

For the experiments in this and the next section, the corpus (dataset 'Wiki33') used to create the reference co-occurrence graph G consisted of 30 randomly selected articles and three intentionally selected articles on the diseases chickenpox, measles and rubella from the English Wikipedia. In Fig. 6.12, it is shown that in general there is a significant distance between the best (the actually chosen one) centroid term and the next 150 potential centroid candidates closest to it.

This experiment has been conducted using 500 randomly selected sentences from the mentioned Wikipedia corpus for which their respective centroid terms have been determined while avoiding a topical bias. The results show that the mean distance from the best centroid to the potential centroid candidates gradually increases, too. Although the standard deviation is relatively large, it stays constant. However, even

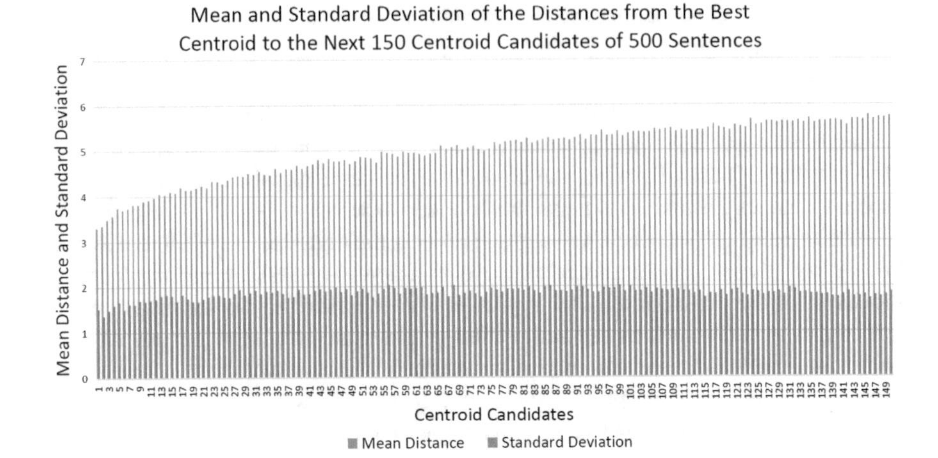

Fig. 6.12 Distances between centroid candidates

when taking this value into account as well, the mean distance in the co-occurrence graph between the best centroid and its e.g. 10 closest centroid candidates is still large enough to come to the conclusion that the determined centroid (its position) is in general the best choice to represent a given textual entity.

At this point, it must be mentioned that the well-known superposition principle may not be applied to text centroids, i.e., if $\chi(D_1)$ and $\chi(D_2)$ are the centroid terms of two pieces of text (or documents) D_1 and D_2, the following usually holds:

$$\chi(D_1 \cup D_2) \neq \chi(\chi(D_1) \cup \chi(D_2)).$$

Figure 6.13 illustrates one respective counterexample. Given the presented graph, the nodes Z_1 and Z_2 are the centroids for the sets of nodes {a, b, c} and {u, v, w,} respectively. These sets could represent the terms contained in two sentences. The following distances between the nodes can be extracted: $d(Z_1, x) = 1$ for $x \in \{a, b, c\}$; $d(Z_2, x) = 1$ for $x \in \{u, v, w\}$; $d(Z_4, Z_1) = d(Z_4, Z_2) = 2$; $d(Z_3, Z_1) = d(Z_3, Z_2) = 3$; $d(Z_3, x) = 2$ for $x \in \{a, b, c, u, v, w\}$.

The centroid of Z_1 and Z_2 is node Z_4 and not Z_3 which is, however, the centroid of all single nodes contained in these sets. In particular, this fact contradicts with the hope to reduce the needed effort to calculate the centroid of a set of documents in an easy manner. It shows once more that there are significant differences between the text centroids and their physical analogon, the centre of mass, due to the discrete character of the co-occurrence graph.

However, experiments have shown, that $\chi(D_1 \cup D_2)$ and $\chi(\chi(D_1) \cup \chi(D_2))$ are not too far from each other. As it can be seen in Table 6.3 for an example using the Wikipedia-article 'Measles', those centroids have an average low distance between 2 and 3 while the maximum distance of two terms in the co-occurrence graph used is 18. The distance between the centroids of all sentence centroids in a fixed section of

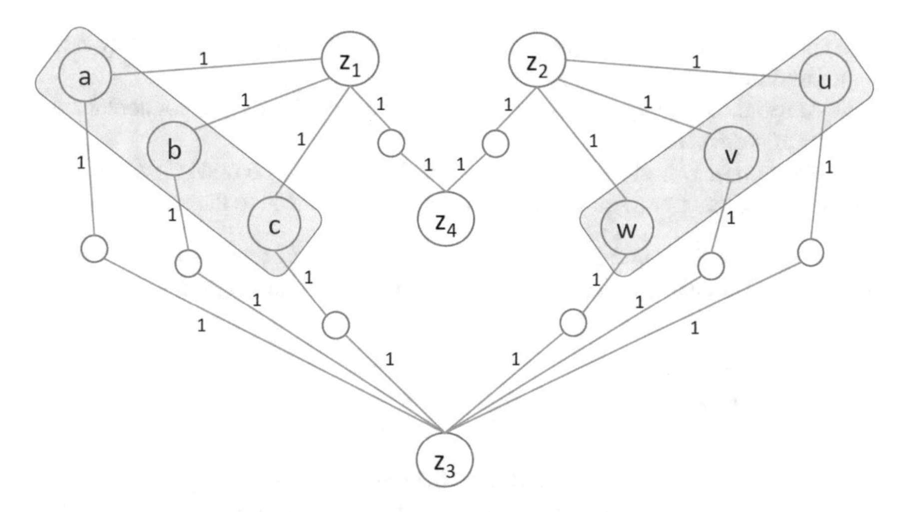

Fig. 6.13 Counterexample for the calculation of centroids

Table 6.3 Distances of specific centroids of sections in the Wikipedia-article 'Measles'

Number of section	Centroid of all sentence centroids in section	Centroid of section	Distance of both centroids
1	Measles	Treatment	3, 63
2	Virus	Measles	2, 31
3	HIV	HIV	0
4	Infection	Diagnosis	3, 40
5	Aid	Measles	3, 24
6	Infection	Risk	2, 52
7	Infection	Symptom	3, 44
8	Net	Prevention	1, 53
9	Malaria	Prevention	2, 23
10	Aid	Interaction	1, 51
11	Research	Health	2, 23

the article and the direct centroid of this section (in this case, its sentence boundaries have not been considered and its terms have been directly used to determine the centroid) is shown for all 11 sections.

Summarising, it can be noted that

- the centroid of a single term is the term itself, the centroid of two terms is usually a node close to the middle of the shortest path between them,
- the centroid is usually not the most frequent or most central term of a document,

- usually, the centroid is uniquely defined although two or more terms may satisfy the condition to be the centroid,
- the centroid of a query or text document can be a term which is not contained in its set of words and
- the centroid of two or more centroid terms is usually not a node on any shortest path among them or a star point with the shortest distance to them.

Nevertheless, finding a representing term to pieces of text also brings with it significant advantages, which shall be discussed in the following section.

6.4.3 Hierarchies of Centroids

Although—differing from semantic approaches—the assigned centroid terms may not necessarily represent any semantic meaning of the given text, they are in each case a formally calculable, well-balanced extract of the words used in the text and their content relations. Nevertheless, centroids may also be used to detect developments regarding content shifts, i.e. subsequent changes in sections, paragraphs or sections of texts may be analysed, where usually classic methods offer only a pairwise comparison of the similarity of text fragments and do not take additional structural information of the given texts into account.

Figures 6.14 and 6.15 show for the two structurally similar Wikipedia-articles 'Measles' and 'Chickenpox' the obtained dendrograms of centroid terms if the centroids of sentences are set in relation with those of paragraphs, sections and the whole documents. The centroids of those text fragments have been calculated by directly taking all terms contained in them into account and not by computing the centroids of the centroids of the respective next lower structural level. As an example for the article 'Measles', 11 sections are contained in it while the first section's centroid is 'treatment', the second section's centroid is 'infection' and so on. Furthermore, the sixth section (on the treatment of measles with the centroid 'risk') contains five paragraphs and with up to 5 sentences in them. For each of those paragraphs and each of the sentences contained in them, the computed centroids are presented, too.

The section on the treatment of chickenpox (also the sixth section) contains in contrast to the article 'Measles' 10 paragraphs. As it can be seen in the depicted lists of sentence centroids for both articles and even on paragraph level, the terms such as 'paracetamol', 'vitamin', 'drug' and 'dipyridamole' are more specific than the ones on section level. The centroid term of the fourth paragraph in the treatment section of the article 'Chickenpox' could not be properly determined ('##unkn##') as the one sentence in this paragraph contains only one term (the noun 'antiviral') which is, in addition to it, not existing in the given co-occurrence graph as it does not co-occur with any other term in the used Wikipedia corpus.

Alternatively, as stated in the previous section, the term 'antiviral' could have been chosen as the centroid term instead. However, it can be noticed from this example and similar cases that the determination of centroid terms is partly difficult and their

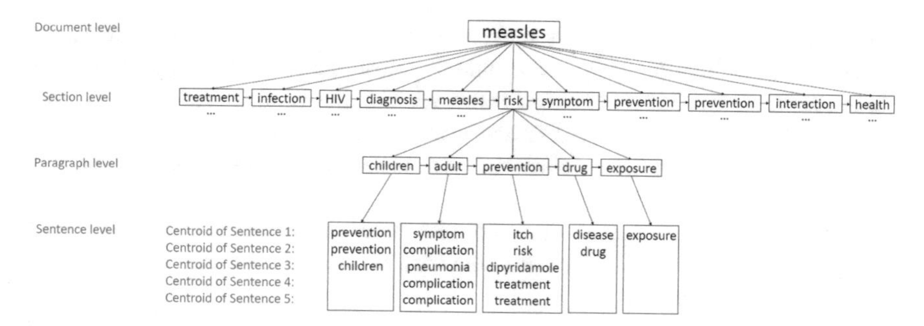

Fig. 6.14 Hierarchy of centroids obtained from sentences, paragraphs, sections and the entire Wikipedia-article 'Measles'

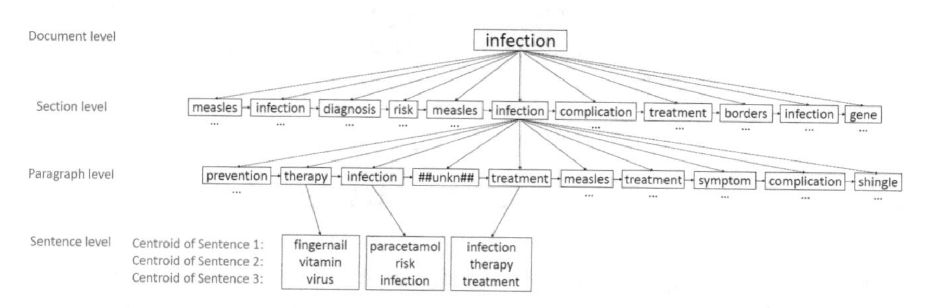

Fig. 6.15 Hierarchy of centroids obtained from sentences, paragraphs, sections and the entire Wikipedia-article 'Chickenpox'

quality is reduced as well when the textual context used for this purpose is small, the co-occurrence graph does not contain required terms or isolated and small clusters in it are addressed. In practice, the handling of exceptional cases of such a kind must be specified.

When comparing both dendrograms (and the centroid terms in them), it is also possible to come to the conclusion that both articles exhibit a similar topical structure. Even on section level, it is recognisable that the articles first deal with the general description of the diseases followed by usual diagnostic methods applied. Then they deal with the treatment of the diseases and finally discuss their epidemiology.

The interesting aspect is not only the decomposition of segments into sub-topics but the traces obtained from the left-to-right sequence of centroid terms and topics on the same level of the tree. From the examples above, it can be concluded that

- the diseases covered in the selected articles are similar,
- the articles exhibit the same structural composition,
- the centroid terms in the lower structural levels are more specific than in the upper levels (the number of terms in the sentences and paragraphs used to calculate them is of course lower and their centroids are determined by smaller, more topically specific contexts),

- a distance calculation of equally-ranked centroids on the same structural level will result in an estimation of how semantically close the respective descriptions (in this case those diseases) are to each other (due to these considerations, the author detected the similarity of two diseases, whose English names are similar but differ from e.g. German ones, i.e. 'Measles' and 'German Measles' (Rubella)) as well as
- a continuous distance check of paths or sequences of section- or paragraph-based centroid terms of very similar documents can show where exactly their semantic or topical differences lie.

Therefore, it is sensible that future cluster building solutions take into account these findings. Also, they present a new direction to compute the centrality of words in a co-occurrence graph in order to find proper generalising terms for contents and to perform further semantic derivations from the position of centroids and their traces in the co-occurrence graph.

6.5 Further Experiments with Centroid Terms

For all of the exemplary experiments discussed in this section, linguistic preprocessing has been applied on the documents to be analysed whereby stop words have been removed and only nouns (in their base form), proper nouns and names have been extracted. In order to build the undirected co-occurrence graph G (as the reference for the centroid distance measure), co-occurrences on sentence level have been extracted. Their significance values have been determined using the Dice coefficient [28]. The particularly used sets of documents will be described in the respective subsections.

6.5.1 Centroid Terms of Wikipedia Articles

As seen in Sect. 6.2, the centroid terms of 30 English Wikipedia articles are actually useful. The corpus used to create the reference co-occurrence graph G consisted of 100 randomly selected articles (including the mentioned 30 ones) from an offline English Wikipedia corpus from http://www.kiwix.org.[5]

In order to extend the previous experiment by also including relationships between terms that do not (or only rarely) co-occur in sentences but significantly co-occur with the same terms (they stand in a paradigmatic relationship to each other), second-order co-occurrences [12] have been extracted as well and additionally used to create

[5]Interested readers may download these datasets (1.3 MB) from: http://www.docanalyser.de/cd-corpora.zip.

the reference co-occurrence graph G. This way, these relevant indirect relationships are taken into account, too.

Table 6.4 compares the centroid terms obtained from both approaches. It can be noted that the centroid terms calculated using the extended reference co-occurrence graph are generally more suitable. However, both approaches returned useful results for further considerations.

Table 6.4 Centroids of 30 Wikipedia articles

Title of Wikipedia article	Centroid term	Centroid term using second-order co-occurrences
Art competitions at the Olympic games	Sculpture	Art
Tay-Sachs disease	Mutation	Therapy
Pythagoras	Pythagoras	Pythagoras
Canberra	Canberra	Capital
Eye (cyclone)	Storm	Storm
Blade runner	Ridley scott	Blade runner
CPU cache	Cache miss	Cache
Rembrandt	Louvre	Paris
Common Unix printing system	Filter	Printer
Psychology	Psychology	Psychology
Religion	Religion	Religion
Universe	Shape	Universe
Mass media	Database	Computer
Rio de Janeiro	Sport	Rio
Stroke	Blood	Brain
Mark Twain	Tale	Twain
Ludwig van Beethoven	Violin	Concerto
Oxyrhynchus	Papyrus	Discovery
Fermi paradox	Civilization	Civilization
Milk	Dairy	Fat
Corinthian war	Sparta	Sparta
Health	Fitness	Health
Tourette syndrome	Tic	Tic
Agriculture	Crop	Agriculture
Finland	Tourism	Government
Malaria	Disease	Malaria
Fiberglass	Fiber	Glass
Continent	Continent	Continent
United States Congress	Senate	Congress
Turquoise	Turquoise	Turquoise

6.5.2 Comparing Similarity Measures

In order to evaluate the effectiveness of the new centroid-based distance measure, its results will be presented and compared to those of the cosine similarity measure while the same 100 online news articles from the German newspaper 'Süddeutsche Zeitung' from the months September, October and November of 2015 have been selected (25 articles from each of the four topical categories 'car', 'travel', 'finance' and 'sports' have been randomly chosen) for this purpose. The following steps have been carried out:

1. As the cosine similarity measure operates on term vectors, the articles' most important terms along with their scores have been determined using the extended PageRank [59] algorithm which has been applied on their own separate (local) co-occurrence graphs (here, another term weighting scheme such as a TF-IDF variant [8] could have been used as well).
2. The cosine similarity measure has then been applied on all pairs of the term vectors. For each article A, a list of the names of the remaining 99 articles has been generated and arranged in descending order according to their cosine similarity to A. An article's A most similar article can therefore be found at the top of this list.
3. Then, the new centroid distance measure to determine the articles' semantic distance has been applied. For this purpose, each article's centroid term has been determined with the help of the co-occurrence graph G using formula 6.6.
4. The pairwise distance between all centroid terms in G of all articles has then been calculated. Additionally, to make the results of the cosine similarity measure and the centroid distance measure comparable, the centroid distance values have been converted into similarity values using formula 6.8.

The exemplary diagram in Fig. 6.16 shows for the reference article ('Abgas-Skandal-Schummel-Motor steckt auch in Audi A4 und A6') its similarity to the 50 most similar articles. The cosine similarity measure was used as the reference measure. Therefore, the most similar article received rank 1 using this measure (blue bars). Although the similarity values of the two measures seem uncorrelated, it is recognisable that especially the articles with a low rank (high similarity) according to the cosine similarity measure are generally regarded as similar by the centroid distance measure, too. In case of Fig. 6.16, the reference article dealt with the car emissions scandal (a heavily discussed topic in late 2015). The articles at the ranks 3 ('Abgas-Affäre—Volkswagen holt fünf Millionen VWs in die Werkstätten'), 7 ('Diesel von Volkswagen—Was VW-Kunden jetzt wissen müssen') and 12 ('Abgas-Skandal—Was auf VW- und Audi-Kunden zukommt') according to the cosine similarity measure have been considered most similar by the centroid distance measure, all of which were indeed related to the reference article. The strongly related articles at the ranks 1, 4, 6 and 9 have been regarded as similar by the centroid distance measure, too. In many experiments, however, the centroid distance measure considered articles as similar although the cosine similarity measure did not.

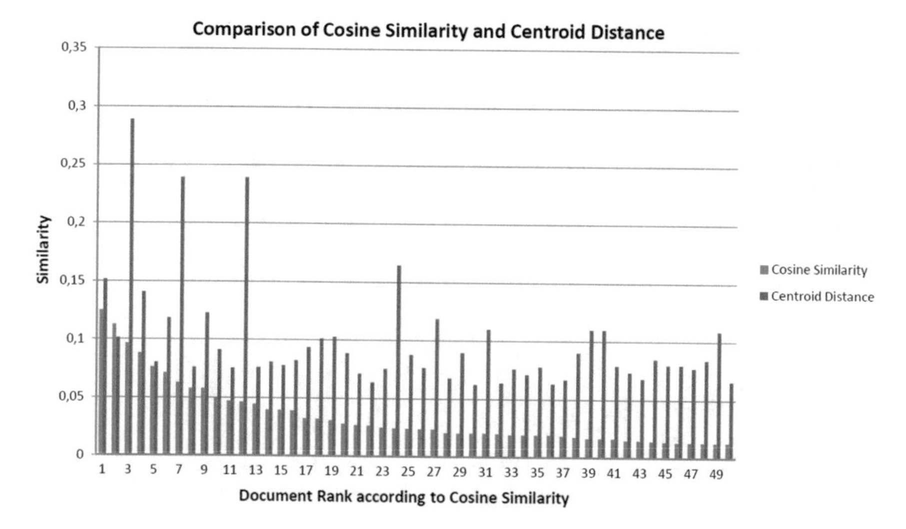

Fig. 6.16 Cosine similarity versus centroid distance (topic: car emissions scandal)

Here, another implicit yet important advantage of the new centroid distance measure becomes obvious: two documents can be regarded as similar although their wording differs (the overlap of their term vectors would be small or even empty and the cosine similarity value would be very low or 0). The article at rank 49 ('Jaguar XF im Fahrbericht—Krallen statt Samtpfoten') is an example for such a case. The centroid distance measure uncovered a topical relationship to the reference article, as both texts are car-related and deal with engine types.

Figure 6.17 depicts another case of this kind: the article with rank 29 received the highest similarity score from the centroid distance measure. A close examination of this case revealed that the centroids of the reference article ('Deutschland—Ausgebucht') and the article in question ('Briefporto—Post lässt schon mal 70-Cent-Marken drucken') are located close to each other in the reference co-occurrence graph. The reference article's main topic was on financial investments in the German hotel business and the article at rank 29 dealt with postage prices of Deutsche Post AG. Both articles also provided short reports on business-related statistics and strategies.

6.5.3 Searching for Text Documents

The previous experiments suggest that the centroid distance measure might be applicable to search for text documents, too. In this sense, one might consider a query as a short text document whose centroid term is determined as described before and the k documents whose centroid terms are closest to the query's centroid term are

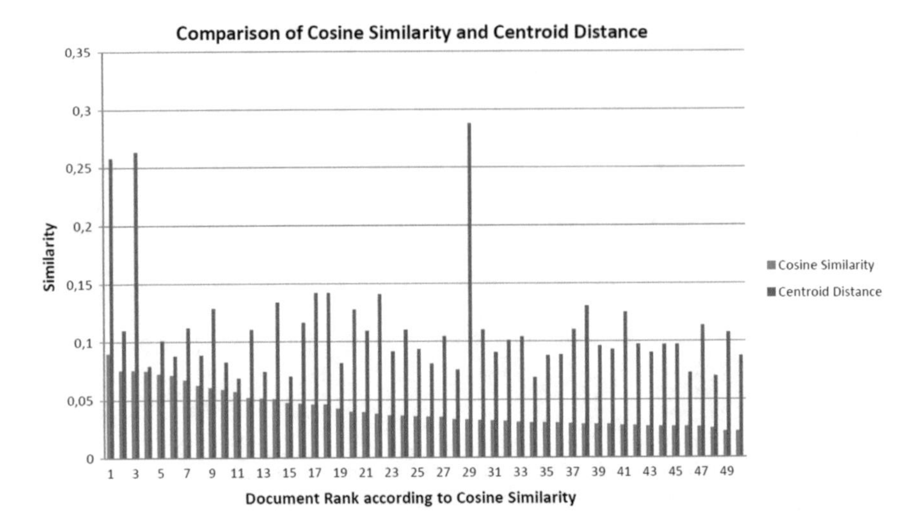

Fig. 6.17 Cosine similarity versus centroid distance measure (topic: business-related statistics and strategies)

returned as matches. These k nearest neighbours are implicitly ranked by the centroid distance measure, too. The best matching document's centroid term has the lowest distance to the query's centroid term.

The following two tables show for two exemplary queries 'VW Audi Abgas' (centroid term: 'Seat' (car brand)) and 'Fußball Geld Fifa' (centroid term: Affäre) their respective top 10 articles from the German newspaper 'Süddeutsche Zeitung' along with their own centroid terms whereby the distances from the queries' centroid terms to all 100 mentioned articles' centroid terms in the co-occurrence graph G have been calculated (Tables 6.5, 6.6).

It can be seen that most of the documents can actually satisfy the information need expressed by the queries. This kind of search will, however, not return exact matches as known from the popular keyword-based web search. Instead, documents will be returned that are in general topically related to the query. As the query and the documents to be searched for are both represented by just one centroid term, an exact match is not possible when applying this approach.

However, this method can still be of use when a preferably large set of topically matching documents is needed. This kind of recall-oriented search is of interest e.g. for people that want to get an overview of a topic or during patent searches when exact query matches might lower the chance of finding possibly relevant documents that nevertheless do not contain all query terms but related terms instead. A typical precision-oriented search would then be harmful.

In order to optimise both recall using the centroid distance measure and also the precision for the k top documents (precision@k), it might be sensible to calculate a combined rank that factors in the rankings of both approaches. Also, it is imaginable to use the centroid distance measure (as a substitute for the Boolean model) to pre-

Table 6.5 Top 10 documents for the query 'VW Audi Abgas' (Seat)

Filename of news article	Centroid term
auto_abgas-skandal-vw-richtet...	Audi
geld_aktien-oeko-fonds-schmeissen-volkswagen-raus	Ethik
auto_bmw-siebener-im-fahrbericht-luxus-laeuft	S-Klasse
auto_abgas-affaere-volkswagen-ruft...	Schadstoffausstoß
auto_abgas-skandal-schummel-motor...	Schadstoffausstoß
geld_briefporto-post-laesst-schon...	Marktanteil
auto_abgas-skandal-was-auf-vw-und-audi...	EA189
auto_abgas-skandal-acht-millionen-vw-autos...	Software
auto_diesel-von-volkswagen-was-vw-kunden...	Motor
auto_abgas-affaere-schmutzige-tricks	Motor

Table 6.6 Top 10 documents for the query 'Fußball Geld Fifa' (Affäre)

Filename of news article	Centroid term
sport_affaere-um-wm-mehr-als-nur-ein-fehler	Fifa
sport_angreifer-von-real-madrid-karim-benzema...	Videoaufnahme
sport_affaere-um-wm-vergabe-zwanziger-schiesst...	Zwanziger
sport_affaere-um-fussball-wm-wie-beckenbauers...	Organisationskomitee
sport_affaere-um-wm-zwanziger-es-gab-eine...	Organisationskomitee
sport_affaere-um-wm-vergabe-zwanziger-legt...	Gerichtsverfahren
sport_affaeren-um-wm-vergaben-die-fifa...	Zahlung
sport_affaere-um-wm-netzer-wirft-zwanziger...	Fifa-Funktionär
geld_ehrenamt-fluechtlingshilfe-die-sich...	Sonderausgabe
sport_affaere-um-wm-wie-zwanziger-niersbach...	Präsident

select those documents that are in a second step ranked according to the cosine similarity measure. Still, other well-known techniques such as expanding queries using highly related and synonymous terms [71] are suitable options to increase recall as well.

Also, in the experiments presented here, mostly topically homogeneous text have been used in order to demonstrate the validity of the centroid distance measure and the role of centroid terms as text representatives. In future experiments, it will be interesting to evaluate the effectiveness of this approach when it is applied on more topically heterogeneous documents.

6.6 Discussion and Fields of Application

The presented approach of using a reference co-occurrence graph to determine the semantic distance of texts is brain-inspired, too. Humans naturally, unconsciously and constantly learn about the entities/objects and their relationships surrounding them and build representations of these perceptions in form of concept maps as well as their terminologies in their minds. New experiences are automatically and in a fraction of a second matched with those previously learned. The same principle is applied when using the centroid distance measure. An incoming text A—regardless of whether it was previously used to construct the co-occurrence graph G or not—whose centroid term shall be found, must at least partially be matched against G. In this sense, G takes on the role of the brain and acts as a global and semantic knowledge base. The only prerequisite is that the graph G must contain enough terms that the incoming text's terms can be matched with. However, it is not necessary to find all of A's terms in G for its—at least rough—topical classification. The human brain does the same. A non-expert reading an online article about biotechnology may not fully understand its terminology, but can at least roughly grasp its content. However, in doing so, this person will gradually learn about the new concepts.

In order to find proper centroid terms for documents whose topical orientation is unknown, it is important to construct the co-occurrence graph G from a preferably large amount of texts covering a wide range of topics. That is why, in the previous experiments, the 100 documents to build the respective corpora have been randomly chosen to create G as a topically well-balanced reference. However, the author assumes that topically oriented corpora can be used as a reference when dealing with documents whose terminology and topical orientation is known in advance, too. This way, the quality of the determined centroid terms should increase as they are expected to be better representations for the individual texts' special topical characteristics. Therefore, a more fine-grained automatic classification of a text should be possible. Further experiments are planned to investigate this assumption.

The bag-of-words model that e.g. the cosine similarity measure solely relies on is used by the centroid-based measure as well, but only to the extend that the entries in the term vectors of documents are used as anchor points in the reference co-occurrence graph G (to 'position' the documents in G) in order to determine their centroid terms. Also, it needs to be pointed out once again that a document's centroid term does not have to occur even once in it. In other words, a centroid term can represent a document, even when it is not mentioned in it.

However, as seen in the experiments, while the cosine similarity measure and the centroid distance measure both often regard especially those documents as similar that actually contain the same terms (their term vectors have a significantly large overlap), one still might argue that both measures can complement each other. The reason for this can be seen in their totally different working principles. While the cosine similarity measure will return a high similarity value for those documents that contain the same terms, the centroid distance measure can uncover a topical relationship between documents even if their wording differs. This is why it might be sensible to combine both approaches in a new measure that factors in the results of both methods. Additional experiments in this regard will be conducted in the future.

Additionally, the herein presented experiments have shown another advantage of the centroid distance measure: its language-independence. It relies on the term relations and term distances in the reference language-specific co-occurrence graph G that has been naturally created using text documents of any language. Although, G is therefore language-specific, the actual determination of the centroid terms as well as application of the centroid distance measure using it is anyhow language-independent.

6.7 Summary

This chapter introduced a graph- and physics-based approach to represent text documents and search queries by single, descriptive terms, called centroid terms. Their properties and suitable fields of applications such as the semantic comparison of text documents have been analysed and discussed in detail. Also, a fast algorithm based on the spreading activation technique has been presented for their efficient determination. Due to their proven characteristics and advantages, centroid terms seem to be perfectly suited for the realisation of the concept of the 'Librarian of the Web'.

In order to show their usefulness for its core tasks, i.e. to topically group documents and to route and forward incoming queries based on semantic considerations, the following chapter introduces respectively designed centroid-based algorithms for library management and document clustering.

Chapter 7
Centroid-Based Library Management and Document Clustering

7.1 Motivation

As mentioned before, the goal is to create a decentralised, librarian-inspired web search engine. In order to do so, a concept is needed to locally classify, sort and catalogue documents as well as to search for them which at the same time avoids the copying of documents (normally carried out by centralised web search engines) and can be implemented as a core module in a P2P-client, the librarian.

For this purpose, it is necessary to take a look at how human librarians could manage their library while it grows. Starting with a few books (or other text documents), the collection will grow until it cannot be managed at one location anymore. Consequently, the set of books must be divided into two subsets and stored separately. Books in one shelf shall have similar or related contents that significantly differ from books in other shelves. Following this idea, characteristic terms in these collections such as names, categories, titles, major subjects (see Sect. 2.1.2 on subject indexing) can be identified by the librarian which are used as the content or index of a card catalogue. Later, the described steps can be applied in the same manner to each sub-collection (Fig. 7.1).

At the same time, the librarian can answer requests by library users with the help of this card catalogue and guide (in a technical sense route) them to bookshelves with appropriate books.

In the following elaborations to derive a proper formal and algorithmic solution for this process, the term librarian refers to the P2P-client that actually performs this library management, document clustering and query routing.

Basically, the process starts with an initially generated root node (librarian) R_1. The refinement level of the librarian k is initially set to $k(R_1) = 1$ and the classification term t of the root node is set to $t(R_1) = NIL$. Links to documents are managed in a local database of each node, denoted by $L(R_k)$. Every node R_k can generate two child nodes, whereby the node identifications R_{2k} and R_{2k+1} are kept on R_k. The local

M. Kubek, *Concepts and Methods for a Librarian of the Web*,
Studies in Big Data 62, https://doi.org/10.1007/978-3-030-23136-1_7

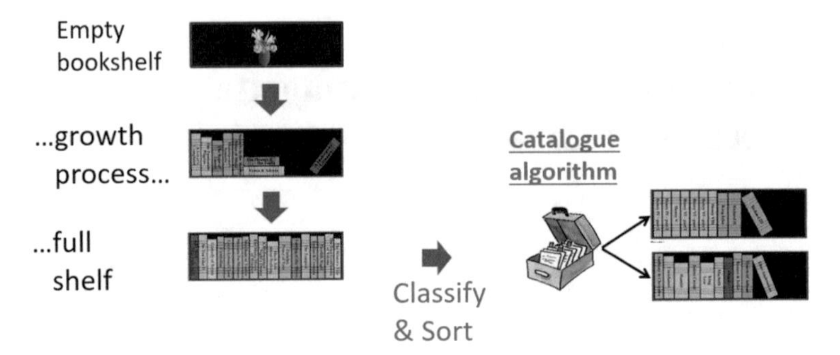

Fig. 7.1 Growth and division of a book collection

co-occurrence graph $G(R_k)$ represents the state of the knowledge of word relations and their distances and corresponds to the respective knowledge of the librarian at this level.

The following description provides an outline of the subsequent algorithms: Initially, the node R_1 at level 1 collects document links until it is full and builds a co-occurrence graph from the terms of the linked documents (Fig. 7.2).

Then, the local co-occurrence graph is partitioned. As the centroid terms of the local documents are terms of this co-occurrence graph, this step automatically generates a document dichotomy (two clusters of document links). Afterwards, two child nodes (refinement level 2) are generated (Fig. 7.3).

These child nodes are assigned document links according to the computed document dichotomy. The co-occurrence graphs on each node of level 2 will be built based on the assigned document links. Level 1 (the co-occurrence graph of node R_1) remains as a means for a rough classification of newly arriving documents/links or queries (Figs. 7.4, 7.5).

The co-occurrence graphs on level 2 are (topically) refined by incoming document links assigned to the respective node. In case the nodes on level 2 are full, the process is repeated.

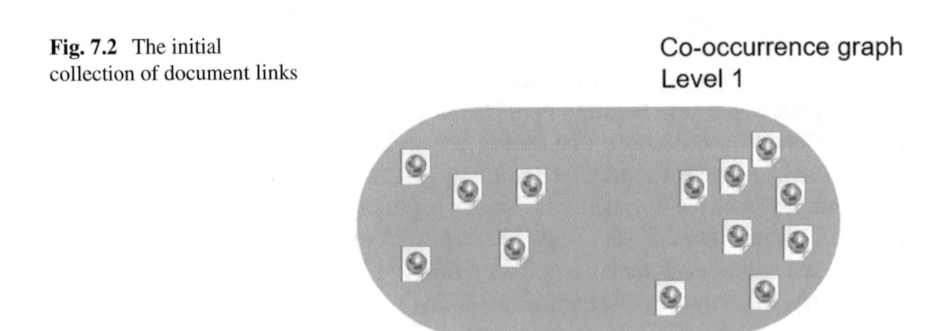

Fig. 7.2 The initial collection of document links

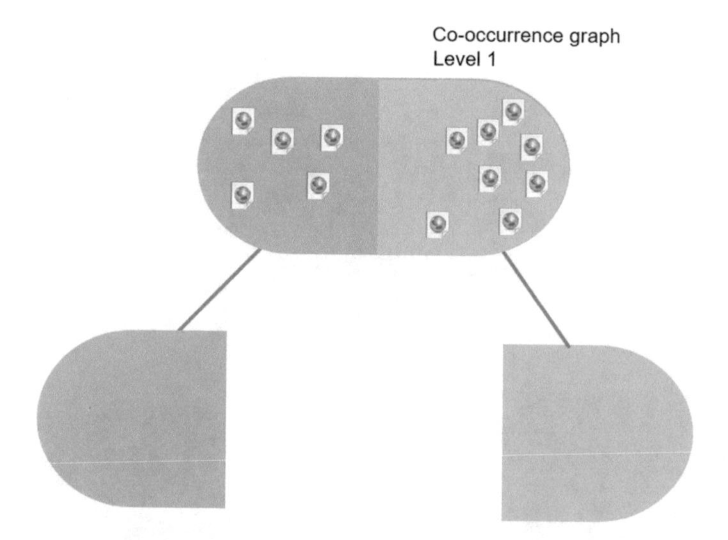

Fig. 7.3 Generating the document dichotomy and child nodes

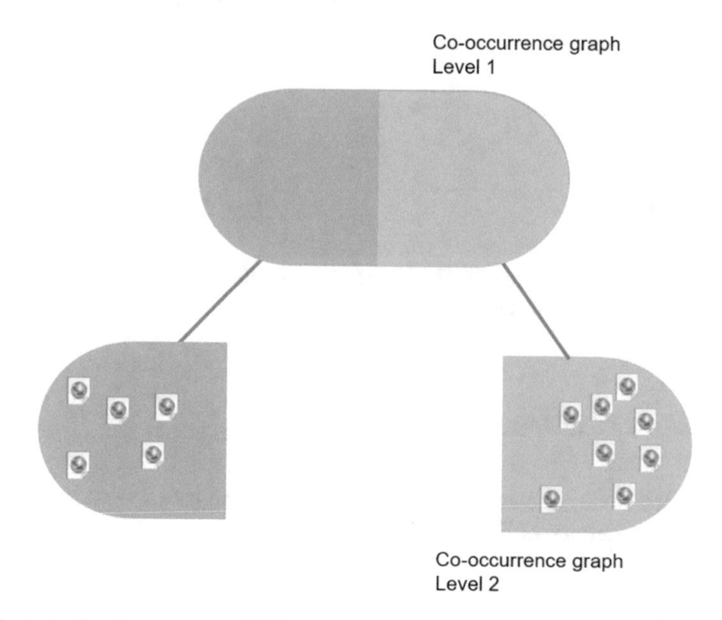

Fig. 7.4 Assigning documents to child nodes

This top-down library management algorithm inherently generates a tree-like document structure (the library) and is locally executed on each node of the P2P-network. As another advantage, the average time complexity to perform searches in this structure is $\mathbf{O}(log(n))$ with n being the number of existing nodes. However, a suitable (likely high-performance) root node must be determined or elected beforehand.

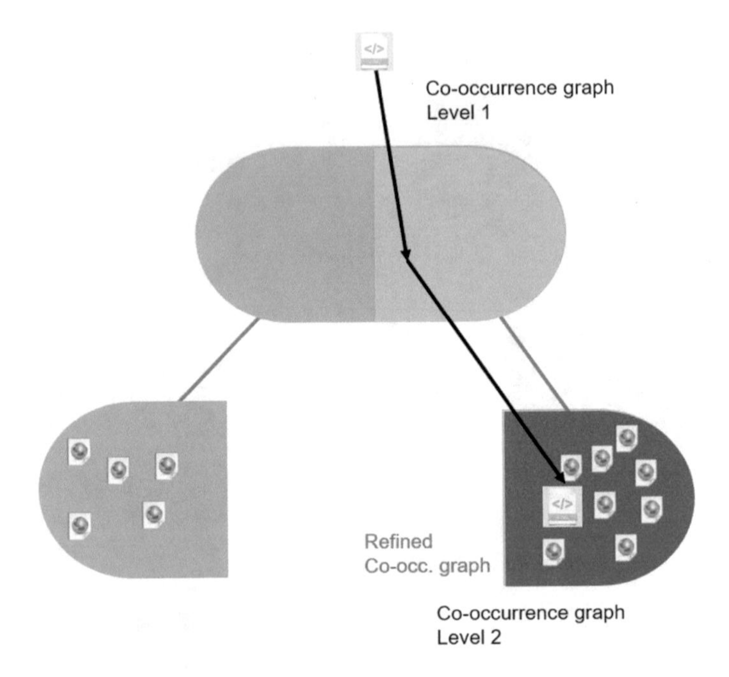

Fig. 7.5 Refinement of co-occurrence graphs

7.2 Algorithms

In this section, the algorithms for the library management and document dichotomy generation are formally described.

7.2.1 Library Management

The mentioned node R_1 starts with the execution of the following algorithm. It is started on any generated child node R_k, too.

1. Receive the initial node data.
2. **REPEAT** //Growth loop
    a. receive and store document links l in $L(R_k)$ on R_k and add their co-occurrences to the local co-occurrence graph $G(R_k)$
    b. receive search queries and answer them using the evaluation of documents addressed by $L(R_k)$.

UNTIL the (designated) memory is full.

3. //Classify and sort
Apply the following cluster algorithms I or II to find a division (dichotomy) of all documents $D(L)$ addressed by links $l_i \in L(R_k)$ into two sets $D_x(l_i), x \in \{1, 2\}$. For later use, define $f_d(T, R_k, D_1, D_2)$ as a function returning the index of either D_1 or D_2 to which any query or text document T is most similar. This index can be determined e.g. using the centroid-based distance, a naive Bayes classifier or any other suitable similarity function.
4. Generate two child nodes of R_k, R_{2k} and R_{2k+1} and send them their respective value for k and t which corresponds to the centroid terms of either one of the dichotomy sets D_1 or D_2. Also send them a copy of $G(R_k)$ for later extension.
5. Move (not copy) the links to every document in D_1 to the node R_{2k} and in D_2 to R_{2k+1}, respectively.
6. **WHILE** //Catalogue and order loop
    a. Receive and calculate for all obtained text links l – i.e. either for incoming, new documents or (sequences of keywords of) search queries $T(l) - x = f_d(T(l), R_k, D_1, D_2)$.
    b. Move/forward the respective document link or search query T to $R_{2k+(x-1)}$.

END.

The selection of the clustering method to determine the dichotomy of documents will significantly influence the effectiveness of this approach. Both of the following algorithms to calculate document dichotomies borrow some ideas from the standard but discrete $k - Means$ clustering algorithm [77] with the parameter k (number of clusters to be generated) set to $k = 2$.

7.2.2 Cluster Algorithm I (Document Dichotomy Generation)

1. Choose two documents $D_1, D_2 \in D(l_i)$, i.e. addressed by an $l_i \in L(R_k)$, such that for their centroid terms t_1 and t_2 in R_k respectively, $d(t_1, t_2) = MAX$. (antipodean documents). If there are several pairs having (almost) the same high distance, choose a pair, for which both centroids have an almost similar, high valence.
Set $D(L) := D(L) \setminus \{D_1, D_2\}$.
2. Randomly choose another document $D_x \in D(L)$ and determine its centroid t_x.
3. If $d(t_x, t_1) \leq d(t_x, t_2)$ in R_k set $c = 1$ and otherwise $c = 2$.
4. Build $D_c := D_c \cup D_x$. In addition, set $D(L) := D(L) \setminus D_x$.
5. While $D(L) \neq \emptyset$, GoTo 2.
6. Determine the new centroid terms $t_c(D_c)$ using R_k for both document sets obtained for $c = 1, 2$, i.e. D_1 and D_2.

In this case, $f_d(T, R_k, D_1, D_2) = 1$, if for a given text or query T with the centroid term t $d(t, t(D_1)) \leq d(t, t(D_2))$ in R_k and otherwise $f_d(T, R_k, D_1, D_2) = 2$. In contrast to the classic k-Means algorithm, the repeated calculation of the updated centroids of both obtained clusters is avoided and carried out only once such that the algorithm runs faster. In order to overcome the (possible) loss of exactness in this sequential process, a modification is made as follows.

7.2.3 Cluster Algorithm II (Document Dichotomy Generation)

1. Choose two documents $D_1, D_2 \in D(l_i)$, i.e. addressed by an $l_i \in L(R_k)$, such that for their centroid terms t_1 and t_2 in R_k respectively, $d(t_1, t_2) = MAX$. If there are several pairs having (almost) the same high distance, choose a pair, for which both centroids have an almost similar, high valence.
   Set $D(L) := D(L) \backslash \{D_1, D_2\}$.
2. Choose another, remaining document $D_x \in D(L)$ such that its centroid t_x is as close as possible to t_1 or t_2.
3. For both document sets D_1 and D_2, i.e. for $i = 1, 2$ calculate the average distance $d(t_x, D_i)$ of the centroid of the newly chosen document D_x to all centroids of the texts $D_{i,1} \dots D_{i,|D_i|}$, which are already assigned to D_1 or D_2 by

$$d(t_x, D_i) = \frac{1}{|D_i|} \sum_{j=1}^{|D_i|} d(t_x, t(D_{i,j})).$$

   If $d(t_x, D_1) \leq d(t_x, D_2)$ set $c = 1$ and otherwise $c = 2$.
4. Build $D_c := D_c \cup D_x$. In addition, set $D(L) := D(L) \backslash D_x$.
5. While $D(L) \neq \emptyset$, GoTo 2.
6. Determine the new centroid terms $t_c(D_c)$ using R_k for both document sets obtained for $c = 1, 2$, i.e. D_1 and D_2.

In this case, the determination of $f_d(T, R_k, D_1, D_2)$ remains the same as presented in cluster algorithm I. In contrast to cluster algorithm I, this algorithm (pre)selects the next document D_x to be assigned to either D_1 or D_2 deterministically. This modification aims to speed up the process to compute meaningful clusters right from the start. Also, the final assignment decision does not only depend on the distance of its centroid term t_x to the centroid terms of the antipodean documents, but on its average distance to all the centroid terms of documents already assigned to D_1 or D_2. This way, the documents in D_1 or D_2 are given a voice in the decision where to assign D_x to.

Since a formal analysis of the described (heuristic) mechanisms seems to be impossible, in the following section, some important properties and clustering results from conducted experiments shall be investigated.

7.3 Cluster Evaluation

In this section, the cluster algorithms introduced above will be evaluated. This evaluation is necessary in order to show that centroid-based clustering algorithms can be helpful to group terms and documents alike and that the overall quality of the generated library structure is high enough to support search processes that involve multiple routing decisions.

7.3.1 Experimental Setup and Preliminary Considerations

For all of the exemplary experiments discussed in this section, linguistic preprocessing has been applied on the documents to be analysed whereby stop words have been removed and only nouns (in their base form), proper nouns and names have been extracted. In order to build the undirected co-occurrence graph G (as the reference for the centroid distance measure), co-occurrences on sentence level have been extracted. Their significance values have been determined using the Dice coefficient. The composition of the used sets of documents will be described in the respective subsections.[1]

The aim of the following preliminary experiments is to show that among the first k documents returned (they have the lowest centroid distance to a reference document) according to the centroid distance measure, a significant amount of documents from the same topical category is found. This experiment has been carried out 100 resp. 200 times for all the documents in the following two datasets (each document in these sets has been used as the reference document). The datasets used consist of online news articles from the German newspaper 'Süddeutsche Zeitung' from the months September, October and November of 2015. Dataset 1.1 contains 100 articles covering the topics 'car' (25), 'money' (25), 'politics' (25) and 'sports' (25); dataset 1.2 contains 200 articles on the same topics with each topic having 50 documents. The articles' categories (tags) have been manually set by their respective authors. On the basis of these assignments (the documents/articles to be processed act as their own gold-standard for evaluation), it is possible to easily find out, how many of the k nearest neighbours (kNN) of a reference document according to the centroid distance measure share its topical assignment. The goal is that this number is as close to $k = 5$ resp. $k = 10$ as possible. For this purpose, the fraction of documents with the same topical tags will be computed.

As an interpretation of Table 7.1, for dataset 1.1 and $k = 5$, the centroid distance measure returned in average 3.9 documents with the reference document's topical assignment first. For the $k = 10$ returned documents first, in average 7.6 documents shared the reference document's tag. The median in both cases is even higher.

---

[1]Interested readers may download these sets (1.3 MB) from: http://www.docanalyser.de/cd-clustering-corpora.zip.

Table 7.1 Average number of documents that share the reference documents' category for their $k = 5$ resp. $k = 10$ most similar documents

Aver. number of doc./median	$k = 5$	$k = 10$
Dataset 1.1	3.9/5	7.6/9
Dataset 1.2	3.9/5	7.5/9

These good values indicate that with the help of the centroid distance measure it is indeed possible to identify semantically close documents. Furthermore, the measure is able to group documents with the same topical tags which is a necessity when building a librarian-like system whose performance should be comparable to human judgement, even when no such assessment is available. This is a requirement that cannot be met by measures relying on the bag-of-words model. The centroid distance measure's application in kNN-based classification systems seems therefore beneficial as well. The findings further suggest that the centroid distance measure is useful in document clustering techniques, too.

That is why, the clustering algorithms I and II apply this measure and are presented and evaluated using a set of experiments in the following subsections. Their effectiveness will be compared to the (usually supervised) naive Bayes algorithm [78] which is generally considered a suitable baseline method for classifying documents into one of two given categories such as 'ham' or 'spam' when dealing with e-mails. As the introduced algorithms' aim is to generate a dichotomy of given text documents, their comparison with the well-known naive Bayes approach is therefore reasonable. While generally accepted evaluation metrics such as entropy and purity [139] will be used to estimate the general quality of the clustering solutions, for one dataset and for all three clustering/classification approaches, the parameters of the resulting clusters will be discussed in detail. In doing so, the effects of the algorithms' properties will be explained and their suitability for the task at hand evaluated.

For these experiments, three datasets consisting of online news articles from the German newspaper 'Süddeutsche Zeitung' from the months September, October and November of 2015 have been compiled:

1. Dataset 2.1 consists of 100 articles covering the topics 'car' (34), 'money' (33) and 'sports' (33).
2. Dataset 2.2 covers 100 articles assigned to the categories 'digital' (33), 'culture' (32) and 'economy' (35).
3. Dataset 2.3 contains 100 articles on the topics 'car' (25), 'money' (25), 'politics' (25) and 'sports' (25).

Although three of the four topical categories of dataset 2.1 are found in dataset 2.3 as well, a different document composition has been chosen. As in the preliminary set of experiments, the articles' categories have been manually set by their respective authors and can be found in the articles' filenames as tags. On the basis of these assignments, it is possible to apply the well-known evaluation metrics entropy and purity. However, these assignments will of course not be taken into account by the clustering/classification algorithms when the actual clustering is carried out.

7.3.2 Experiment 1: Clustering Using Antipodean Documents

In the first experiment discussed herein, algorithm I is iteratively applied in two rounds on the clustering dataset 2.1: first on the dataset's initial cluster (root) and then again on its two subclusters (child clusters) created. This way, a cluster hierarchy (binary tree of clusters) is obtained. In Fig. 7.6, this hierarchy is shown. The clusters contain values for the following parameters (if they have been calculated, otherwise N/A):

1. the number of documents/articles in the cluster
2. the number of terms in the cluster
3. the cluster radius (while relying on the respective co-occurrence graph G of the father cluster)
4. the fraction of topics in the set of documents
5. the centroid term of the first document (antipodean document) in the cluster

Moreover, the distance between the centroid terms of the antipodean documents in two child clusters and the intersection of two child clusters (number of terms they have in common) are given in the cluster hierarchy.

It is recognisable that already after the first iteration, the two clusters exhibit dominant topics. E.g. 28 articles with car-related contents are grouped in the first cluster (from left to right). The second cluster contains—in contrast—altogether 51 articles on the topics 'money' and 'sports'. This grouping is not suprising as many of the sports-related articles dealt with the 2015 FIFA corruption case. In the second

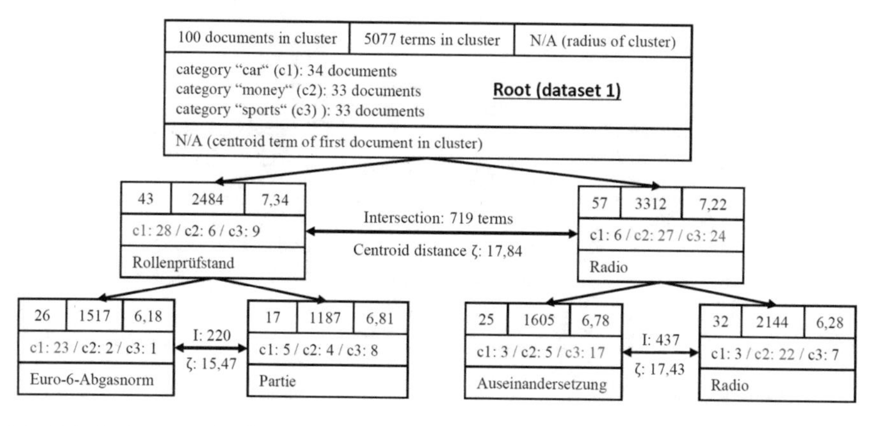

Fig. 7.6 Basic approach: clustering using antipodes

iteration, this cluster is split again (while using its own documents to construct the co-occurrence graph G) creating one cluster with the dominant topic 'sports' (17 articles) and another one with the dominant topic 'money' (22 documents). Also, the recognisable topical imbalance of the first two clusters is a sign that the clustering solution actually works. The documents with the topics 'money' and 'sports' are semantically closer to each other than to the car-related documents. If this unbalance would not occur, it would mean that the clustering would not work properly.

As it can be seen in the clusters, the centroid terms of the first documents (the antipodean documents) in them are very distant from each other in the respectively used co-occurrence graph G (e.g. the distance of the centroid term 'Rollenprüfstand' to the term 'Radio' is 17.84). This means, that their topics differ as well. It is also logical that the documents in D_x to be assigned to one of the two clusters in each iteration share a topical relatedness with one of the two antipodean documents. In all cases, the cluster radii are much smaller than the distance between the centroid terms of the antipodean documents in G.

The calculated cumulated entropy for the generated four clusters in the second iteration is 0.61 and the cumulated purity of these clusters is 0.70. This reasonable result shows that the basic cluster algorithm I (of course depending on the number and the size of the documents to be grouped) is able to return useful clusters after only a few clustering iterations.

7.3.3 Experiment 2: Clustering Using Centroid Terms

In the second experiment, algorithm II is also applied in two rounds on the clustering dataset 2.1: first on the dataset's initial cluster (root) and then again on its two subclusters created. In this case, a cluster hierarchy is obtained as well. Figure 7.7 presents this hierarchy after two clustering iterations. Its structure follows the one given in the first experiment.

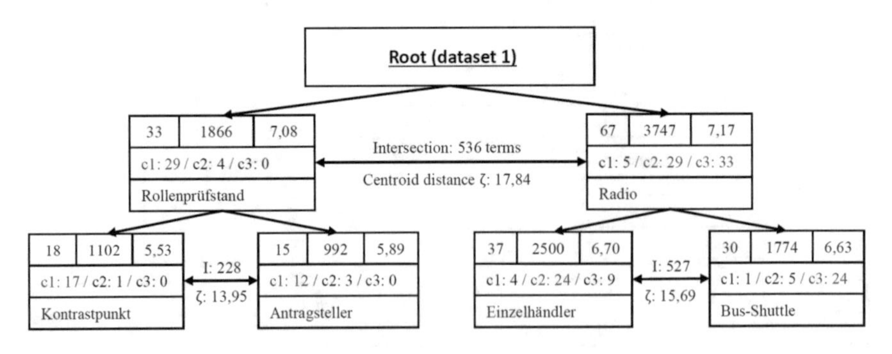

Fig. 7.7 Centroid-based clustering

In contrast to the experiment 1, the first cluster contains one more document from the category 'car' and no document from the category 'sports'. The second cluster contains with 62 documents from the categories 'money' and 'sports' 11 documents more than the second cluster from experiment 1. Here, the topical imbalance of the first two clusters is recognisable, too. Also, the clusters generated in the second iteration exhibit a clear topical orientation. The cluster radii are much smaller than the distance between the centroid terms of the antipodean documents in G, too.

The calculated cumulated entropy for the generated four clusters in the second iteration is 0.55 and the cumulated purity of these clusters is 0.77. This result is even better than the one from algorithm I and shows that it is sensible to not solely base the classification decision on the distance of a document's centroid term to one of the antipodean documents' centroid terms, but to take into account its average distance to all of the already added centroid terms found in one of the two clusters, too.

7.3.4 *Experiment 3: Clustering Using Naive Bayes*

In experiment 3, the well-known naive Bayes algorithm [78] is applied to iteratively and hierarchically group the documents of dataset 2.1. This supervised algorithm is usually applied to classify documents into two categories such as 'ham' or 'spam' when incoming e-mails need to be filtered. It is often regarded as a baseline method when comparing classification techniques. Therefore, it makes sense to use this algorithm to classify the documents of the mentioned datasets into one of two groups in each classification step. However, in order to correctly classify unseen documents, a classifier using the naive Bayes algorithm needs to be trained with particular sets of documents from the categories of interest. Small sets are usually sufficient. For this purpose, the automatically determined antipodean documents are used as this training set. Based on the features (terms) in these documents, the naive Bayes algorithm can determine the probabilities of whether a document from the set D_x rather belongs to either one cluster or the other. A newly classified document (its features) is then automatically taken into account to train the classifier for the next documents from D_x to be classified/clustered.

Here, however, a problem arises (especially when only a few documents from D_x have been classified so far): it might be that the majority of those documents will be assigned to only one category, which is undesirable. The classification probabilities might be shifted in favour of exactly this category and new documents might be wrongly classified, too. Therefore, a preselection of the next document to be classified is applied before the actual naive Bayes classification is executed. In this step, the desired category is determined in an alternating way and the best suited document from the (remaining) non-empty set D_x is selected based on the shortest distance to the centroid term of the antipodean document of this specific category. The aim of this approach is that the two clusters grow at almost the same rate. This approach resembles the human (i.e. manual and supervised) (pre-)classification of documents, before the classifier is trained based on their features. As an example, an e-mail

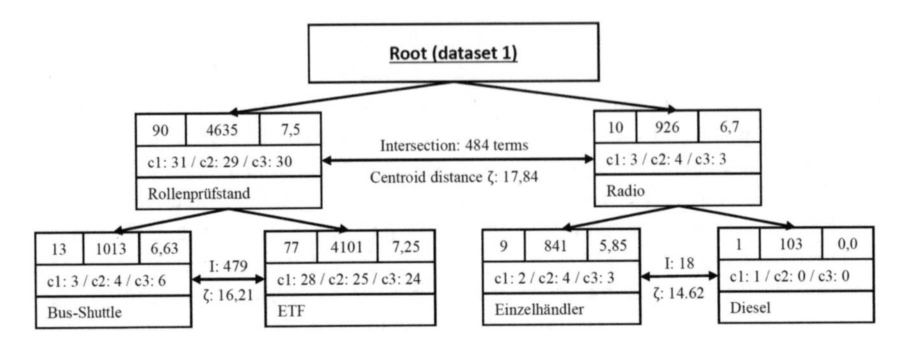

Fig. 7.8 Clustering using a Naive Bayes classifier

can usually be instantly and without much effort categorised by a human reader. The same principle is applied here, only that in the case at hand this preselection is carried out fully automatically. In this setting, the naive Bayes algorithm is applied in an unsupervised way. Although the mentioned constant growth of the two clusters is practically unreachable due to the datasets' characteristics, this preselection is however sensible as described before.

Also in this experiment, a cluster hierarchy is obtained. Figure 7.8 presents this hierarchy after two clustering iterations in the same fashion as in the previous experiments. In contrast to the first and second experiment, the first generated cluster contains almost all documents from the root cluster. Its second child cluster contains with 77 documents covering all topics still a large fraction of all given documents. The remaining clusters are practically unusable due to their topical mixture and the low number of documents (one cluster is actually made up of the initial antipodean document) they contain. Also, no dominant topics can be identified when analysing the clusters in the hierarchy.

This bad result is also reflected in the values of the cumulated entropy and purity. The calculated entropy for the generated four clusters in the second iteration is 0.98 and the purity of these clusters is 0.39. The reasons for this result can be found in both the dataset used and the working principle of the naive Bayes classifier. First, the dataset's documents have many terms in common. The (sub)topic 'money' is found in the documents of the categories 'sports' and 'car', too. For instance, many car-related documents dealt with the car emissions scandal and financial penalties for the car companies involved and a lot of sports-related documents covered the 2015 FIFA corruption case. Second, based on just two training documents (the antipodean documents), the naive Bayes algorithm was not able to separate the given dataset properly. As this algorithm does not make use of the term relations in the respective co-occurrence graphs G, only the features (terms) in the documents, which are supposed to be independent, determine the classification probabilities.

In order to improve the algorithm's clustering performance, an idea was to increase the training set (although the naive Bayes algorithm can be trained on small sets, too). For this purpose and in a small modification of the presented setting, not only the antipodean documents have been initially put into the respective two child clusters, but their two closest documents (in terms of their centroid terms' distance in the co-occurrence graph G), too. This way, any child cluster initially contained three documents. However, even with this modification, the quality of the clusters did not increase.

7.4 General Evaluation and Discussion

The algorithms I, II and the naive Bayes algorithm have been applied on the datasets 2.2 and 2.3, too. Table 7.2 presents the values for the cumulated entropy (a value near 0 is wished for) and cumulated purity (a value near 1 is desired) for all algorithm/dataset combinations. In all these cases, it is to be expected that a topical imbalance occurs in the clusters as seen in the experiments 1 and 2. As datasets 2.1 and 2.2 cover three topical categories, it can therefore be assumed that after already two clustering iterations, a clear topical separation should be visible (in case the algorithms work properly). For dataset 2.3, however, a clear topical separation should be visible after at most three iterations as it contains documents of four topical categories. Therefore, for datasets 2.1 and 2.2, the cumulated entropy and purity values have been computed after two clustering iterations, whereas for dataset 2.3, these values have been determined after three iterations.

As previously described, algorithm II performs best on dataset 2.1. For the datasets 2 and 3, this picture does not change as well. Also, in all cases, algorithm I achieved better entropy and purity values than the naive Bayes algorithm. This shows that the introduced algorithms I and II can clearly outperform the naive Bayes algorithm, even after a small number of cluster iterations.

In this regard, it is sensible to ask, when to start and stop the clustering process? In case of the librarian, the process will be started only once when the node's 'bookshelf' is full and is not carried out again afterwards by this node (but its child nodes). This parameter therefore depends on the hardware resources available at each node running the librarian. In future contributions, this hardware-dependent parameter estimation will be thematised.

The main goal of the herein presented experiments and results was to demonstrate that centroid terms of text documents and their distances are well-suited for clustering purposes. It could be seen in the simulations that the decision base (the co-occurrence

Table 7.2 Entropy and purity of the obtained clusters

Entropy (E)/Purity (P):	Alg. I	Alg. II	Naive Bayes
Dataset 2.1 (100 doc./3 topics)	E = 0.61 P = 0.70	**E = 0.55 P = 0.77**	E = 0.98 P = 0.39
Dataset 2.2 (100 doc./3 topics)	E = 0.65 P = 0.69	**E = 0.46 P = 0.82**	E = 0.97 P = 0.37
Dataset 2.3 (100 doc./4 topics)	E = 0.41 P = 0.76	**E = 0.35 P = 0.82**	E = 0.49 P = 0.62

graph G) to put a document in either one of the two possible clusters is reduced in size in each clustering iteration. This means that with a growing number of iterations the probability to make the right decision (correctly order the stored documents) is reduced, too. Even so, it was shown that hierarchic clustering based on centroid distances works satisfying.

In case of the presented librarian (the real implementation following the library management algorithm), however, the decision base would not shrink as the clustering process would be carried out once after a node's local 'bookshelf'/memory is full. The local co-occurrence graph G is then handed down to its two child nodes and is at this place continuously specialised by incoming (more topically focused) documents. At the same time, the area of the original graph G for which a child node is primarily responsible (represented by the terms of its local document repository) is reduced in size. This decision area is therefore more specialised than the whole graph of the father node. Thus, child nodes can make sharp classification decisions on incoming documents of their topical specialisation, yet are able to classify documents of a different topical orientations correctly. After the generation of child nodes, the librarian only acts as a semantic router for incoming document links and search queries as described in step 6 of the library management algorithm.

7.5 Summary

A concept for the decentralised, content-based document management and search of text documents, subsumed under the term 'library management', has been presented in this chapter. At its core, it comprises two new hierarchic text clustering methods that compute semantic distances of documents using their centroid terms and generate clusters of comparably high quality. Also, search queries are processed by the same approach as they are regarded as 'small documents' which can be represented by centroid terms as well.

The output of the introduced library management algorithm, which is designed to locally run on the peers (the librarian) of a P2P-network, is a tree-like document structure (the actual library) which is efficiently searchable. Its construction principle follows the procedures of human librarians to classify and catalogue books. In both processes, search and library construction alike, centroid terms play the role of the small cards of a card catalogue to either guide users to relevant documents or to classify and catalogue incoming documents correctly.

In order to show that these centroid terms are of major benefit in interactive search processes, too, the following chapter addresses this assumption and complements the basic findings from Chap. 6.

Chapter 8
Centroid-Based Search Support

8.1 Evaluating Search Results and Query Classification

The following experiments have been conducted in order to show that centroid terms can successfully be applied to search for documents in the clusters of the generated library structure, i.e. in the local document sets of the created (child) nodes. Especially, it is investigated, if it is possible to return matching documents by solely relying on the distances of text- and query-representing centroid terms in reference co-occurrences graphs, if queries can be topically classified by centroid terms and if the direct neighbours of centroids as well as the second or third best centroid candidates could be better representatives for the search subject? In doing so, the initial results presented in Sect. 6.5.3 are greatly extended.

8.1.1 Experimental Setup

In the following experiments, the two datasets 'Corpus-100' and 'Corpus-1000' have been used to construct the co-occurrence graphs for the determination of centroid terms. They consist of either 100 or 1000 topically classified (topical tags assigned by their authors) online news articles from the German newspapers 'Süddeutsche Zeitung' and 'Die Welt'. Each dataset consists of 25 respectively 250 randomly chosen articles from the four topical categories 'car' (German tags: 'Auto', 'Motor'), 'money' (German tags: 'Geld', 'Finanzen'), 'politics' (German tag: 'Politik') and 'sports' (German tag: 'Sport').[1]

In order to build the (undirected) reference co-occurrence graphs, linguistic preprocessing has been applied on these documents whereby sentences have been extracted, stop words have been removed and only nouns (in their base form),

[1]Interested readers may download the used datasets and analysis results (10.1 MB) from: http://www.docanalyser.de/search-corpora.zip.

M. Kubek, *Concepts and Methods for a Librarian of the Web*,
Studies in Big Data 62, https://doi.org/10.1007/978-3-030-23136-1_8

proper nouns and names have been considered. Based on these preparatory works, co-occurrences on sentence level have been extracted. Their significance values have been determined using the Dice coefficient [28] and its reciprocal has been used to represent the distance between the terms involved. These values and the extracted terms are persistently saved in an embedded Neo4j (https://neo4j.com/) graph database using its property-value store provided for all nodes (represent the terms) and relationships (represent the co-occurrences and their significances).

The analysed articles stem from the same datasets and their representing centroid terms have been determined using these co-occurrence graphs and persistently stored afterwards. Furthermore, each article's nouns, proper nouns and names have been stored in an inverted index in order to be able to answer Boolean queries as well. All results are provided along with the mentioned datasets.

8.1.2 Experiment 1a: Centroid-Based Search

In the first set of experiments presented here, the goal is to show that for automatically generated and topically homogeneous queries consisting of three terms with a low diversity (see Sect. 6.3.2 and the following section) (neighbouring terms in the co-occurrence graph of dataset 'Corpus-1000') from the four mentioned topics, the first 10 (Top-10) returned documents are actually relevant. In order to test the applicability of centroid terms in this process, the centroids of these queries have been determined along with their distances to the centroid terms of the articles in the co-occurrence graph of dataset 'Corpus-1000'. Then, for each query, the articles have been listed in ascending order by this distance. Furthermore, these queries have been sent to Google as well. It has been made sure that the generated queries make sense, i.e. that humans would actually formulate them.

While it is not directly possibly to compare the results of the much smaller dataset 'Corpus-1000' with the huge result set of Google, a general evaluation of the retrieval quality—especially in an interactive setting—is possible. Therefore and because of the shortcomings of the well-known evaluation metrics recall and precision, for each query, the set of the Top-10 results of both approaches (centroid-based approach and Google) have been assigned marks by five test persons while applying the German school grading scale. The marks 1 (very good), 2 (good), 3 (satisfactory), 4 (sufficient) and 5 (not sufficient) were possible.

The evaluation results of 20 randomly generated queries (five of each topic) are presented in Table 8.1. In it, the average and rounded marks of all five test persons are presented for each query. As it is possible to see in these results, generally, the Top-10 results obtained from the centroid-based approach are mostly relevant. This also implies that mostly useful documents from the correct topical orientation have been returned first from the set of all possible 1000 articles. Although, Google's results were generally more relevant, its much larger knowledge base is certainly one reason for this outcome. The overall, average mark for the centroid-based approach was 2, 0 (good) and 1, 4 for Google (very good). However, a closer look at the results reveals

Table 8.1 Evaluation of centroid-based (CB) and Google's search results

Query no.	Query (topic: car)	Centroid	Mark CB	Mark Google
1	Bauzeit, VW-Konzern, Millionenseller	Bauzeit	3	2
2	Luftreinhalteplan, Software-Nachrüstung, Stuttgart	Luftreinhalteplan	2	1
3	Roller, Hersteller, Gefühl	Roller	3	2
4	Abgasrückführung, Grenze, Spritverbrauch	Abgasrückführung	2	1
5	Ruß, Abgasreinigung, Rohrleitung	Ruß	2	4
Query no.	Query (topic: money)	Centroid	Mark CB	Mark Google
6	Loft, Monatsmiete, Quadratmeter	Loft	2	1
7	mTan, Auftrag, Verfahren	mTan	3	1
8	Mischfond, IMC, Investmentfond	Mischfond	2	1
9	Stoppkurs, Prozent, Zertifikat	Aktie	2	1
10	Verbraucher, Sicherheit, Sparer	Sparer	1	1
Query no.	Query (topic: politics)	Centroid	Mark CB	Mark Google
11	Flüchtlingsheim, Gesundheit, Albaner	Flüchtlingsheim	2	1
12	Asylpolitik, Region, Bundesjustizminister	Asylpolitik	2	2
13	Migration, Monat, Untersuchung	Migration	3	2
14	BAMF, BA, Mitarbeiter	BAMF	3	2
15	Flüchtlingslager, Damaskus, Europaparlament	Flüchtlingslager	1	1
Query no.	Query (topic: sports)	Centroid	Mark CB	Mark Google
16	Stadion, Fan, Lust	Stadion	2	1
17	Halbzeit, Spiel, Abpfiff	Halbzeit	1	1
18	Schiedsrichter, DFB, Hinweis	DFB	2	1
19	Spieltag, Weise, Bundesliga-Live	Spieltag	1	1
20	Stadion, Verhaltenshinweise, Mannschaft	Stadion	2	1
		Average mark:	**2, 0**	**1, 4**

that Google was not able to interpret all queries correctly. As an example, for the car-related query 5, Google's results were largely related to heating technology and industrial engines and only two relevant results had a relation to car emissions.

Another observation is that the sports-related terms seem to have a higher discriminative power as the centroid-based approach returned in almost all cases relevant results for this topic. The other topics contain more terms that can be found in many documents of different topical orientations. Especially money-related terms can be found in the documents of all topics. For instance, in the car-related documents that mostly deal with the car emissions scandal in 2015, financial penalties are often discussed. Therefore, these topics' queries and their respective centroids have a lower discriminative power leading to a slightly higher number of irrelevant search results.

8.1.3 Experiment 1b: Boolean Retrieval and Partial Matching

Also, Boolean retrieval and partial matching has been applied on these 20 queries while using the inverted index of the dataset "Corpus-1000". These queries have been interpreted as conjunctive ones, which means that the goal was to find documents that contain all query terms. The results obtained from these tests are presented in Table 8.2 which shows the number of exact matches and the number of partial matches for each query. In only 10 cases, a document matching all terms could be found. As another noteworthy example, for query 8, only one exactly matching article and only one partially matching article could be found although 250 money-related (among

Table 8.2 Results of Boolean retrieval and partial matching

Query no.	Exact matches	Partial matches
1	0	15
2	1	59
3	1	167
4	1	159
5	1	9
6	1	43
7	1	86
8	1	1
9	1	316
10	1	91
11	0	45
12	0	75
13	0	276
14	0	72
15	0	30
16	0	116
17	0	47
18	1	69
19	0	93
20	0	145

them many relevant stock market news) articles were available. Also, queries 1 and 5 returned only a minor fraction of the relevant documents to be found. However, in most of the cases, partial matching returned at least an acceptable number of documents. However, the general finding for these retrieval approaches is that they are unable to identify a large quantity of relevant results. A low recall is to expected when applying them. Even so, the documents they return are mostly relevant as the query terms actually appear in them (at least partially).

8.1.4 Experiment 2: Classification of Query Topics

The second set of experiments investigate whether the queries' topical orientation can be correctly identified by the centroid-based approach. In order to evaluate this automatic classification, for each query, the number of the Top-10 returned documents that match its topical orientation are counted and the average number as well as the median of topical matches have been determined based on these numbers afterwards. At first, this measurement has been carried out for the 20 low-diversity queries listed above. Table 8.3 presents the results of these measurements. As it is to be seen, for all topics, the topical matches outweigh the mismatches.

In accordance with the previous experiment, the sports-related queries generally returned more topically matching articles than the queries from the other topics. Generally, it can be deducted that the centroid-based search approach is for the most part able to correctly identify a query's topical orientation, especially when it is a low-diversity one.

In order to evaluate the influence of the query diversity, the same experiment has been carried out for topically oriented queries with unspecified diversity as well. That means that the queries' terms—in contract to the previous experiments—do not necessarily have to be direct neighbours in the used reference co-occurrence graph generated from the dataset 'Corpus-1000'. Instead, the queries' terms have been automatically and randomly selected from a particular topic of interest. As a consequence, those queries might not always make sense for humans, but a mixture of low- and high-diversity queries has been obtained this way. In this experiment, for each of the four mentioned topics, 25 queries (altogether 200 queries) have been automatically generated.

Table 8.3 Detection of topics of low-diversity queries

Topic	Average no. of topical matches	Median of topical matches
Car	6.4	6
Money	7.4	8
Politics	6.6	6
Sports	8.8	8

Table 8.4 Detection of topics of queries with unspecified diversity

Topic	Average no. of topical matches	Median of topical matches
Car	4.3	4
Money	4.26	5
Politics	6.06	7
Sports	6.42	8

As Table 8.4 shows, the average number and the median of topical matches expectedly decreased. The reason for this result is that due to their generally higher diversity, the queries' centroid terms are more distant to the query terms as well and therefore often can not be clearly assigned to one or the other topic. Usually, those centroid terms are general ones (e.g. the German word 'Jahr', Engl. year) that have many connections to terms of all other topics in the co-occurrence graph. If such kind of terms are determined as centroid terms, the Top-10 documents returned are mostly from diverse topics as well. However, even in this experiment, the sports-related queries returned the best results. Their median of topical matches even did not decrease at all.

8.1.5 Experiment 3: Centroid Candidates

While the previous results indicate that centroid terms can in fact be useful in search or classification processes, there is still room for improvement. As an example, the finally chosen centroid term C which has by definition the shortest average distance to the terms of a query—these terms might stem from a user query or a text document alike—in the used co-occurrence graph, might not be the best selection to represent it. Instead, the usage of the second or third best centroid candidates might be an even better choice. The following experiments investigate this hypothesis.

Furthermore, the direct neighbours of C on the shortest paths to these candidates might be better descriptors than the sole centroid term C. Last but not least, the experiments' aim is to show that by using larger (and specially analysed) corpora to create the co-occurrence graph, the query interpretation is improved, too. For this purpose, the previously introduced datasets 'Corpus-100' and 'Corpus-1000' have been used. In order to obtain even more term relationships than the standard sentence-level co-occurrence analysis yields, second-order co-occurrences [12] have been extracted from the dataset 'Corpus-1000' as well. Using this technique, relations between terms can be extracted even when they do not co-occur in sentences originally. Differing from the previous experiments, the queries consist of the 25 most frequent terms of the articles to be analysed from which the respective centroid term C and the next best centroid candidates CC_i with $i \geq 2$ are computed.

The following Tables 8.5, 8.6, 8.7 and 8.8 present for four randomly chosen articles from the German newspapers 'Süddeutsche Zeitung' and 'Die Welt' their respectively obtained analysis results. In the tables, the co-occurrence graph numbers refer to the datasets used (1 means 'Corpus-100', 2 means 'Corpus-1000' and 3 means 'Corpus-1000' with second-order co-occurrence extraction). In the following elaborations, the second best centroid candidate is referred to CC_2. The next three centroid candidates are referred to in an analogous manner. The direct neighbours of centroid term C on the shortest path to CC_i are respectively referred to as (neighbouring terms) NT_i with $i \geq 2$.

The results consistently show that by using the larger co-occurrence graphs 2 and 3 the finally obtained centroid terms C better reflect the individual documents' topics. As an example, in Table 8.5, a car-related document has been analysed. When using the small dataset 'Corpus-100', the centroid term 'Modell' (Engl. model) is chosen, which is a rather general term. However, for the co-occurrence graphs 2 and 3, the centroid term 'VW' is returned. This term better reflects the article's content as it deals with the 2015 car emissions scandal which mainly revolves around the German car brand Volkswagen (VW). Likewise, these results confirm that the method to calculate centroid terms is inherently corpus-dependent.

Furthermore, in all test cases, the next best centroid candidates CC_i provide more useful insight into the document's topical focus and have for the largest part a more discriminative power than the actually chosen centroid term. For instance, while in case of the money-related article (Table 8.6), the actual centroid term 'Prozent' (Engl. percent) obtained when using the co-occurrence graphs 1 and 2 has a rather low discriminative power, the centroid candidates 'Euro', 'Bank' (Engl. bank) and 'Geld' (Engl. money) better reflect the article's topical orientation. Therefore, these terms should not be neglected in classifying queries as well. In this example, the comparatively best centroid term (and candidate) is 'Aktie' (Engl. stock) which is returned when using the co-occurrence graph 3 as the article mainly deals with stock market investments.

Similarly, the direct neighbours NT_i of the centroid term C on the shortest path to the centroid candidates CC_i often make the analysed document's topic more explicit than the centroid term itself. In case of the politics-related article (it deals with the European migrant crisis from 2015) used in Table 8.7, topically fitting neighbouring terms of C are for instance Flüchtling (Engl. refugee) or Grenze (Engl. border) which describe the article's focus very well.

Also, it is noteworthy that in many cases (and especially when applying the much denser co-occurrence graph 3) the terms CC_i and NT_i are the same. This observation confirms that the nodes of the next best centroid candidates CC_i are in fact very close or even adjacent to the node of the centroid term C. Usually, only one hop from it is needed to reach them. This even suggests that it is possible to approximate further centroid candidates by simply visiting C's neighbouring nodes (preferably the ones that are closest to the analysed document's terms). By using this approach, there would be no need for computing them explicitly.

Table 8.5 Centroid candidates of the car-related document auto_vw-abgas-skandal-millionen-audis-von-abgasaffaere-betroffen.txt

Basic parameters		Centroid candidates CC_i and direct neighbours NT_i of C on the shortest path to CC_i							
Co-occ. graph	Centroid C	CC_2	NT_2	CC_3	NT_3	CC_4	NT_4	CC_5	NT_5
1	Modell	Fahrzeug	Hersteller	VW	Passat	Million	Hersteller	Konzern	TDI
2	VW	Volkswagen	Marke	Dieselmotor	Dieselmotor	Fahrzeug	Fahrzeug	USA	USA
3	VW	Volkswagen	Volkswagen	Software	Software	Fahrzeug	Fahrzeug	Modell	Modell

Table 8.6 Centroid candidates of the money-related document geld_geldwerkstatt-verlust-verhindern.txt

Basic parameters		Centr. Cand. CC_i and Dir. Neighb. NT_i of C on the shortest path to CC_i							
Co-occ. graph	Centroid C	CC_2	NT_2	CC_3	NT_3	CC_4	NT_4	CC_5	NT_5
1	Prozent	Monat	Euro	Frage	Laufzeit	Euro	Euro	Million	Euro
2	Prozent	Euro	Euro	Jahr	Jahr	Bank	Bank	Geld	Geld
3	Aktie	Prozent	Prozent	Euro	Milliarde	Jahr	Milliarde	Milliarde	Milliarde

Table 8.7 Centroid candidates of the politics-related document politik_fluechtlinge-in-europa-so-wurde-budapest-wien-muenchen-zur-hauptroute-fuer-fluechtlinge.txt

Basic parameters		Centroid candidates CC_i and direct neighbours NT_i of C on the shortest path to CC_i							
Co-occ. graph	Centroid C	CC_2	NT_2	CC_3	NT_3	CC_4	NT_4	CC_5	NT_5
1	Ungarn	Deutschland	Deutschland	Grenze	Grenze	Flüchtling	Flüchtling	Million	Flüchtling
2	Flüchtling	Mensch	Mensch	Deutschland	Deutschland	Land	Land	Euro	Million
3	Flüchtling	Ungarn	Ungarn	Grenze	Grenze	Mensch	Mensch	Deutschland	Deutschland

Table 8.8 Centroid candidates of the sports-related document sport_bundesliga-thomas-tuchels-erfolgreiche-rueckkehr.txt

Basic parameters		Centroid candidates CC_i and direct neighbours NT_i of C on the shortest path to CC_i							
Co-occ. graph	Centroid C	CC_2	NT_2	CC_3	NT_3	CC_4	NT_4	CC_5	NT_5
1	Platz	Europa	Dreier	Sonntag	Sonntag	Bayern	Punkt	Zuschauer	Halbzeit
2	Bayern	Trainer	Dortmund	Punkt	Punkt	Mannschaft	Mannschaft	Partie	Punkt
3	Dortmund	League	League	Punkt	Punkt	Bayern	Bayern	Spiel	Spiel

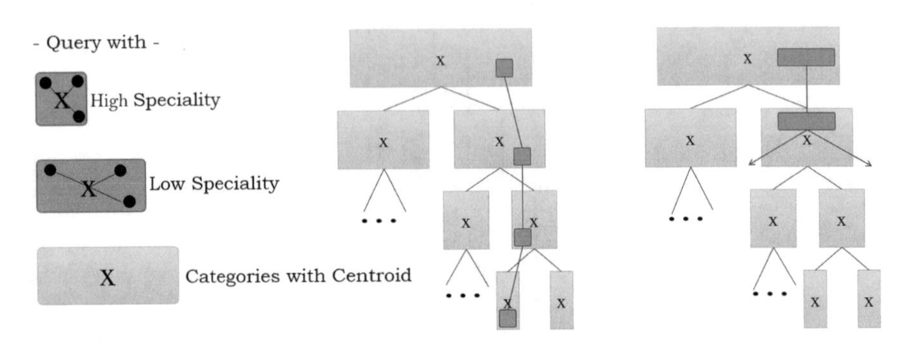

Fig. 8.1 Routing of search queries depending on their speciality

8.2 Supporting Interactive Search

As already indicated in Sect. 6.3.2, a qualitative evaluation of queries by their **diversity** and **speciality** is possible.

As a first application for these measures, the routing of queries arriving at any nodes in decentralised tree-like document structures (the grey-shaded parts in Fig. 8.1) following the presented approach in the previous chapter shall be named. Starting from the query's root node, search queries need to be forwarded to the son node whose centroid has the minimal distance to the centroid of the query or whose area of topical responsibility contains the respective query centroid.

As long as the speciality of the query is high (Fig. 8.1), the decision on where to forward it to is unique and this process can be carried out easily. Problems occur in case of low speciality values of the search query as it may be positively answered by documents offered by the leaves of different subtrees, i.e. the routing is not uniquely defined.

In this case, several strategies can be applied:

1. the most relevant alternative (while exclusively depending on the distance or membership of the centroid term) is chosen, what may, however, reduce recall,
2. the query is duplicated and forwarded to both son nodes of the recent position, what may significantly increase traffic volumes especially in case of multiple applications.
3. an interaction is initiated, whereby
    - additional keywords are presented to the user,
    - context- or session-related information is used to suggest further keywords,
    - centroids and further keywords of the son nodes in the hierarchical structure are suggested to the user as additional criteria and
    - neighbours of centroids or second and third best centroid candidates of the incoming query (or document) are used to solve ambiguity problems

    in order to finally make a more precise decision on where the request must be routed to.

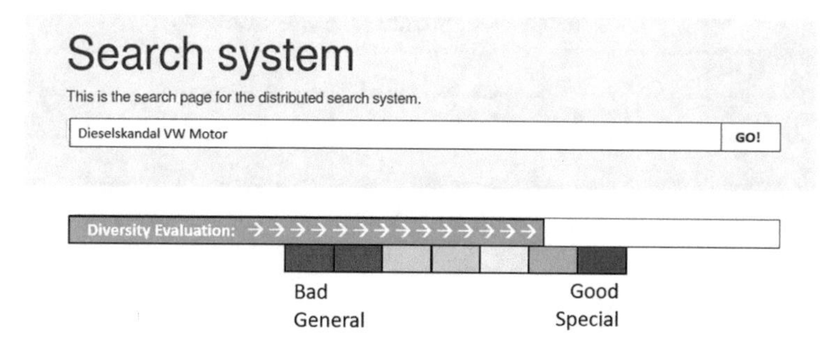

Fig. 8.2 Evaluation of search queries by their diversity

A further field of application for these measures in interactive search sessions can be derived directly from their definitions: They can immediately evaluate a query's (topical) broadness and indicate whether a user should reformulate it. As an example, queries with high diversity values (possibly normalised using the average distance in the co-occurrence graph) stand for unspecialised or comprehensive requests with a huge number of possible, mostly non-relevant results. These queries should therefore be reformulated. Small diversity values represent highly specialised requests, for which usually fewer, very well matching search results may be offered. Respective queries should be accepted as they are.

As Fig. 8.2 exemplary shows, an interactive search system can apply the diversity measure to visually indicate the current query's quality. On top of this evaluation, the system can provide additional guidance and support for users by offering relevant keywords that likely will narrow down the search space and thus return more relevant results and at the same time reduce the number of (potentially) matching documents. Of course, this approach will be particularly useful in precision-oriented web search sessions. Also, this kind of instant feedback can be given in real-time and even before a query is actually sent to and processed by the searcher component (see Sect. 3.2) of a web search engine.

8.3 User-Based Document Ranking

For the most part, content-related aspects have been discussed so far to measure the relevance of documents to user queries as well as to rank them accordingly. In this section, a novel method to rank documents according to their user access rates in the generated tree-like library structure is presented. The following considerations motivate its conception.

As mentioned before, Google and other web search engines usually evaluate the topological position of website in the web graph as a criterion for ranking purposes. Algorithms like PageRank [92] and HITS [55] developed for this purpose. Besides requiring a huge amount of computational resources, those methods are unable to

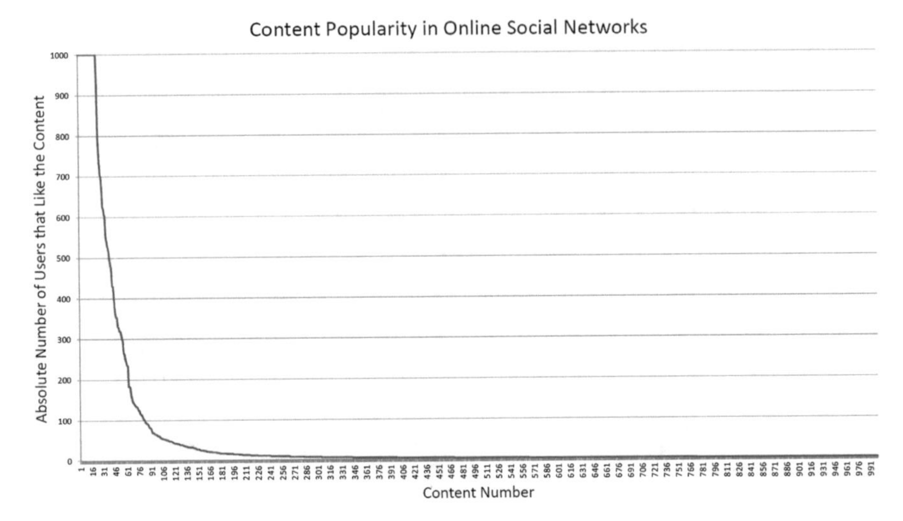

Fig. 8.3 Distribution of content popularity

consider any user feedback for doing so. Moreover, due to reasons of data protection, many users may deny any observations to generate such kind of feedback. However, contents should be ranked highly when they are frequently accessed by users, i.e. when there is high demand for them (Fig. 8.3).

Besides the absolute number of page hits [86] in a given time period, the user's dwell time [4, 50] on a webpage (during which it is actively inspected in the browser) may be a significant feedback for the ranking of such documents, too. For this approach, the client-sided web browsers just have to transfer the visiting time of a webpage to any independent third party which carries out the ranking of webpages by cumulating their observed visiting times in a given time period. This avoids manipulations, since user clients do not have—differing from the accessed hosting servers—any interest in manipulating those information and frequent outliers can be easily detected and removed.

Also, content popularity in the web is power-law distributed [6], i.e. only a few contents have high access rates while most of the other contents are only very seldom visited. Furthermore, this effect does not seem to depend on particular network characteristics such as friendship relations in online social networks. Extraordinary popular contents (and therefore frequently accessed ones) will appear even without word-of-mouth-propagation or other recommendation techniques [23].

Consequently, a significantly lower amount of storage is needed to persistently save frequently accessed contents in the given tree-like document structure. This relation corresponds to node numbers on different levels of this structure. The root node (and connected child nodes) of such a structure might be a high-performance machine (super node or peer) and should therefore be responsible to serve those frequently accessed contents such as text documents in favour of less demanded ones.

This way, the number of times a query needs to routed and forwarded is reduced, too.

In the next subsections, a new approach for a user-based document ranking will be presented while taking the above given facts and previous results into account.

8.3.1 Conceptual Approach

It is intended to use the natural analogon of the air bladder of fish for this kind of document ranking.

The swim bladder of a fish can be contracted and released by muscles. Therefore the volume of a fish is decreased or increased while its weight remains the same. The larger or smaller amount of water displaced by the fish results in a smaller or higher buoyancy of it and consequently in a deeper or more shallow swimming level under the water.

For the document evaluation, the following model may be derived:

- Document links are stored in airtight sealed, elastic balloons, representing the fish and respectively its air bladder.
- The balloons containing the document links are embedded in a deep, liquid (water) environment.
- Since all air bladders are empty in the beginning, they are situated initially on the ground of the given water environment.
- Every access to a document adds air to the balloon of the corresponding document, the amount may depend either

  1. just on the number of user hits,
  2. on the overall duration of the user access or
  3. on the duration of the user access in relation to the size of the document.

- The balloons may loose air by osmotic processes over time.
- The depth of the balloons corresponds to the level in which the each balloon is in a perfect balance between weight and buoyancy.

If this model is used, documents swimming closer to the surface of the liquid pool are those which are accessed more often by the users or have a higher dwell time. As a result, starting an investigation from the surface, more important documents will be found first.

8.3.2 Algorithm Design

This model needs to be adapted to the special needs of the assumed decentralised computer environment and document library structure. This comes along with some—rather artificial—limitations and constructs:

1. The original document data base structure is a binary, tree obtained from a hierarchical clustering process as described in Chap. 7 and usually managed in a decentralised P2P-system like the one formed by the WebEngine-clients (see the next chapter for their implementation).
2. Document links and therefore the described balloon models can not be freely moved but are assigned to nodes of a tree.
3. Consequently, the depth is discretised, i.e. every node represents a depth interval.
4. Nodes of the same tree level shall (must) represent the same depth interval.
5. The nodes have a limited capacity to store document links.
6. Only discrete time intervals will be considered.

For scalability purposes of the system built, some more assumptions shall be derived from content access rates by users.

- It is known that the user interest in content is power-law distributed. The number of documents with a high popularity and therefore high access rates is exponentially decreasing. As a consequence, the capacity of the nodes in the upper tree levels shall be sufficient to store all documents of higher access rates.
- The number of documents managed may grow and therefore the number of tree levels may be changed.
- In particular, parameters of the systems must be dynamically adapted, in the initial phase as well as in case capacity borders are reached.
- In order to simplify the calculation process, buoyancy-weight-balances in a given liquid shall not be calculated, instead the absolute amount of contained gas in the balloon shall be directly used to derive the corresponding depth (node level), in which the respective object is in balance.

Following the above said, the subsequent assumptions and definitions are derived and outlined:

1. A balloon, corresponding to a document $d \in D$ contains at a given time t the amount $V(d, t)$ of gas.
2. At $t = 0$, let $V(d, 0) = 0 \; \forall d \in D$.
3. Every access to the document d adds a ΔV of gas to the balloon, whereby ΔV can be a function of the number of access, the (summarised) duration of viewing the document in a discrete time interval or weighted by the size of the document.
4. In every (discrete) time interval of t, air disappears from each balloon. This is modelled by an exponential function $\Delta V(d, t+1) = V(d, t)e^{-\lambda}$.
5. Consequently, the amount of gas in a given balloon d can be calculated by

$$V(d, t+1) = V(d, t)e^{-\lambda} + \Delta V$$

Furthermore, the following **Rules and Settings** are needed:

1. The document tree has a depth of N containing $2^N - 1$ nodes $n_1, n_2, \ldots, n_{2^N-1}$, whereby n_1 shall be the root of the tree. Consequently, a node n_x at level l $(1 \leq l \leq N)$ has the two son nodes n_{2x} and n_{2x+1} at level $l + 1$. (Note that the

root node n_1 is on level 1, the leaves of a complete, balanced tree reside on level N.)
2. Every node n_x contains an upper and a lower volume border $o(n_x)$ and $u(n_x)$ describing the volume interval of gas, which the document balloons shall have to be in balance on this node.
3. If a node n_x at level l of the tree has the two son nodes n_{2x} and n_{2x+1} at level $l+1$, $u(n_x) = o(n_{2x}) = o(n_{2x+1})$ must be set.
4. Furthermore, $o(n_x) > u(n_x)$ must be ensured.
5. Therefore, for a node n_x on level l of the tree, $u(n_x) \geq \Delta u \cdot (N - l)$ shall be ensured.
6. Initially, for all leaf nodes n_y, $\forall y = 2^{(n-1)} \dots (2^n - 1)$, $u(n_y) = 0$ shall be permanently set (i.e. without a possibility for adaptations).
7. For the root node n_1, $o(n_1) = \infty$ is permanently set.

Initially, all documents $d \in D$ are assigned to leaf nodes of the tree, following the hierarchical clustering generated by the librarian's algorithm described in Chap. 7. Here, the function $h(d)$ returns the leaf node, a document has been initially assigned to.

On every node x of the tree and at each time step t, a set of documents $d \in D(x)$, $D(x) \subseteq D$ is now managed as described by the following **Document Management Algorithm**:

1. Calculate $\forall d \in D(x)$

$$V(d, t+1) = V(d, t)e^{-\lambda} + \Delta V \tag{8.1}$$

2. If $V(d, t+1) \geq o(n_x)$, move d to the father node $int(x/2)$ of x.
3. If $V(d, t+1) < u(n_x)$, move d to n_{2x}, iff there is in the tree a path from n_{2x} to $h(d)$ and otherwise move it to n_{2x+1}
4. Wait until $t = t + 1$ and GoTo 1.

Adaptation

So far, document links $d \in D$ are moved up and down in the tree structure by just depending on their $V(d, t)$. Since every node has only a limited capacity C to keep document links, the lower and upper bounds (borders) $u(n_x)$ and $o(n_x)$ must be subject to an adaptation process to avoid empty or overfull nodes especially in the first levels of the tree close to its root.

Initially, the leaves will contain up to C document links. If approximately $\frac{C}{2}$ links will be allowed to move upwards in the tree, a power law distribution of popular documents will be achieved while node capacity borders can be properly taken into account.

Consequently, all nodes shall be responsible to accept and store $\frac{C}{2}$ document links and in case of deviations of more than $+/-\epsilon$ to adapt their lower border $u(n_x)$ by a given, fixed quantum Δu. This must automatically result in changes for the upper borders of both son nodes n_{2x} and n_{2x+1}.

This **Border Adaptation Process** may be described (in the decentralised P2P-version) as follows for every node n_x:

1. Initialize the $u(n_x)$ and $o(n_x)$ depending on the level $1 \dots N$ and following the above given *Rules and Settings*. Set time step $t = 0$.
2. **LOOP**:
   Set time step $t := t + 1$. Update for all documents the links assigned by $V(d, t)$. Move the links as described in the *Document Management Algorithm* if needed.
3. Check, if document links shall be received from father or son nodes and receive those document links.
4. Check, if adaptations have been sent from the father node and adapt $o(n_x)$ immediately if indicated. If by doing so $o(n_x) \leq u(n_x)$ is obtained, set $u(n_x) := o(n_x) - \Delta u$ and communicate the made adaptations to both son nodes n_{2x} and n_{2x+1}.
5. Adapt $u(n_x)$ for node n_x on level l of the tree, if
   - $l < N$,
   - for at least T time steps no adaptations have been made by the node itself (an exception is only allowed, if the number of documents on the node n_x is critically close to C) and
   - for the current overall number of all document links $c(n_x)$ on n_x

$$|c(n_x) - \frac{C}{2}| > \epsilon$$

   by

$$\begin{aligned} u(n_x) :=& MIN(MAX((u(n_x)+ \\ & \Delta u \cdot sign(c(n_x) - \frac{C}{2})), \\ & (\Delta u \cdot (N - l))), o(n_x)). \end{aligned} \tag{8.2}$$

6. Send possibly made adaptations to both son nodes n_{2x} and n_{2x+1}.
7. GoTo LOOP.

As this user-based ranking is applied on the tree-like document structure (the library) by its managing nodes, it is implicitly carried out within the structure of the search space. At the same time, it is fair, since it takes into account the hits of all users in an equal manner. To visualise the concept, Fig. 8.4 shows the fictional result of the applied ranking process. Only one very popular document has been moved up to the root of the tree; two documents are less popular but have been assigned to nodes at lower node levels. Less popular documents remain at the leave nodes.

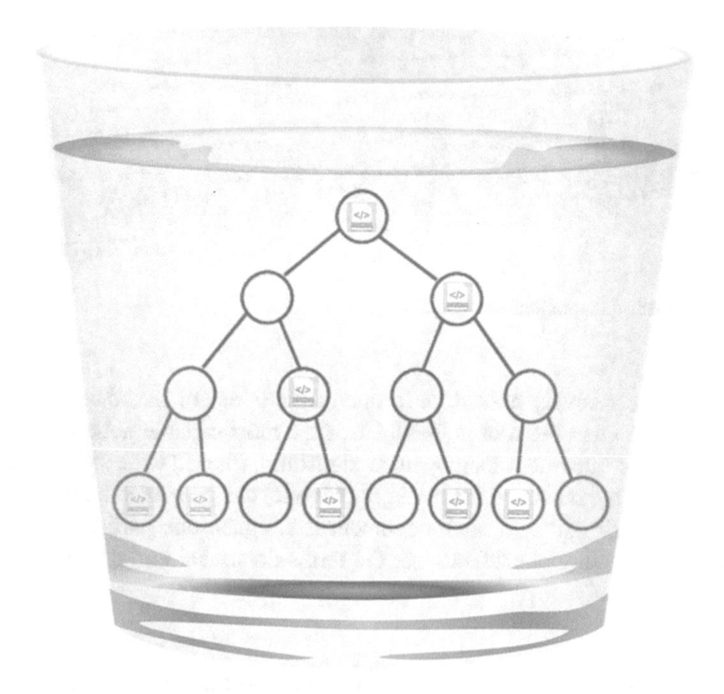

Fig. 8.4 Situation after applied ranking

8.3.3 Experimental Evaluation

In this section, an experimental evaluation of the document management and border adaptation algorithms depending on a varying number of documents (respectively their links) is provided. In the initialisation phase, the used document links are assigned to all leaf nodes available such that they are equally distributed and (almost) all leaf nodes store the same number of document links. Furthermore, in order to initialise the simulations, the following parameters have been uniformly set:

- number of node levels: 5 (31 nodes in tree, 16 leaf nodes)
- node capacity C: 5
- T: 3
- Δu: 2
- ϵ: 5

Additionally, each document link d is assigned a value $V(d)$ in the range of 0 to 50, representing the number of times it has been clicked by users. The values of $V(d)$ are power-law distributed among all documents links. A higher value indicates a document link which is often accessed by users (popular content) and a lower value a less popular content. However, in the simulations, this value does not change after it has been assigned. During the simulation runs, a highly popular content should

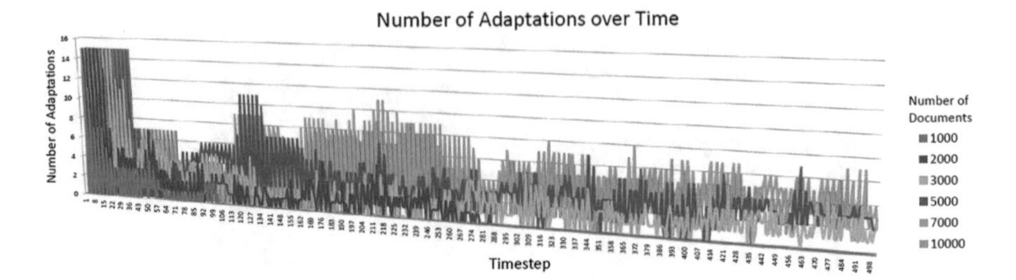

Fig. 8.5 Number of adaptations over time

therefore be successively passed on to one of the nodes of the lower levels in the tree (possibly to even the root node which is the most capable node as discussed). In contrast to the document management algorithm, in each time step, at most two document links (the most and the least popular one) can be passed on to other nodes in the tree by each node. As each document is assigned one particular link and to simplify the following considerations, the terms document and document link are used interchangeably.

8.3.3.1 Experiment 1: Convergence Time Depending on Number of Documents

In the first experiment presented in this context, the goal is to show that even when the number of documents (their links) is growing, convergence is reached, i.e. the number of border adaptations does not or only slightly change after a certain number of (consecutive) time steps. In Fig. 8.5, it can be seen for all six test runs with 1000, 2000, 3000, 5000, 7000 and 10000 documents, the number of adaptations (cumulated over all nodes) is drastically decreasing in the first 50 time steps. Full convergence has been reached after 100 time steps when 1000 documents have been used and after 285 time steps for 2000 documents. Remarkably, even for larger document sets, only a very low number of adaptations (with respect to the overall number of documents dealt with) occurs after about 500 time steps. This result clearly shows that the presented border adaptation algorithm can be used to reach a stable value for the nodes' upper and lower borders after a comparably low number of time steps.

8.3.3.2 Experiment 2: Number of Documents per Node Level

The goal of the second experiment is to show that the number of documents per node level over time develops as anticipated. For this experiment, 1000 documents have been initially distributed among the leaf nodes. As Fig. 8.6 confirms, the number of documents at the leaf nodes slowly but steadily decreases. During the first time steps, the number of documents at node level 4 is increasing. This is not surprising

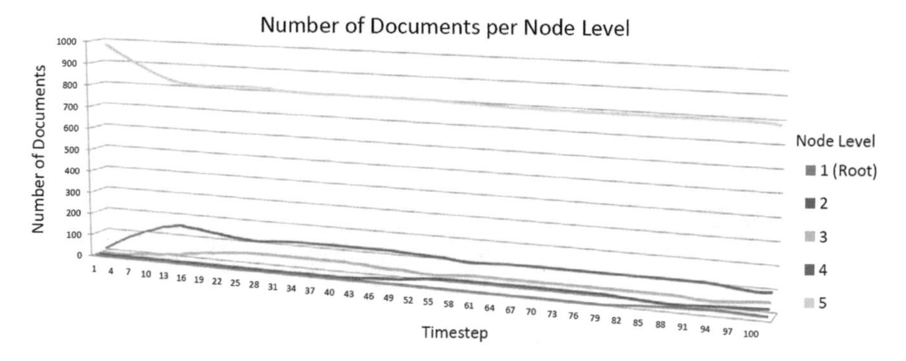

Fig. 8.6 Number of documents per node level

as at that time, this level will receive most of the documents from the leaf nodes. Later on, some of these documents with a high value of $V(d)$ are then passed on to the lower node levels 3 and 2. After about 80 time steps, also the root node will receive documents from its two son nodes. After the simulation run, it can be noted that most of the document links (788) can still be found on the leaf nodes as they are rarely accessed. A much lower number of more popular document links have been moved to the nodes at the levels 4 (94), 3 (60) and 2 (39) while the root node only contains the few most popular ones (19). According to the border adaptation algorithm, adaptations are not allowed for the leaf nodes. Therefore, no adaptations occur at node level 5. The power-law distributed popularity value $V(d)$ is therefore reflected in the document links' position in the tree. This result demonstrates that the document management algorithm works correctly.

8.3.3.3 Experiment 3: Number of Adaptations per Node Level

It shall be investigated here at which node level and at which time border adaptations occur. This experiment has been conducted using 1000 documents for this purpose as well. Figure 8.7 clearly shows that most of the adaptations occur at the lower time steps and especially at the node levels 4 and 3. After 60 time steps, the major part of the adaptations has been carried out. If adaptations occur after this time, as it is to be seen for the root node, these adaptations lead to further adaptations at the lower node levels (after document links have been passed to one of the son nodes). However, the number of these induced adaptations is comparably low. Also, at this time, most of the document links have already been moved to suitable nodes in the tree. Therefore, the number of this kind of adaptations is low.

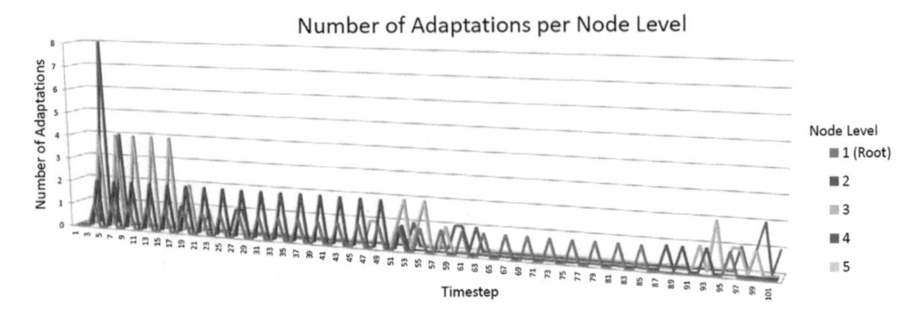

Fig. 8.7 Number of adaptations per node level

8.3.3.4 Further Observations

The previous results showed that an adaptive and converging document ranking in a specially structured document library is possible, even when the document collection is growing. As mentioned, the algorithm parameters C, T, Δu and ϵ have been assigned fixed values in this simulation setup. However, further experiments have shown that the modification of these parameters influences the convergence time and the number of border adaptations as well. As an example, a large value of C (depending on the number of documents in the collection) will reduce the overall number of document links passed on in each time step as they reside longer on a particular node until it is (over)full. As a consequence, the convergence time is increased as well. A large value of T has the same effect. A low value of ϵ causes a node to perform border adaptations more frequently. It will become more sensitive to short-term peaks of high load (many document links received lately). It will therefore try to reduce its load as fast as possible by first increasing the value of $u(n_x)$ and second by sending excess document links to connected nodes. These frequent link movements have a negative effect on the convergence time. A high value of ϵ reduces the number of these adaptations and link movements as well as in doing so the convergence time. However, as a negative effect, it is likely that the nodes at the lower tree levels will stay empty or receive only a very small number of document links.

8.4 General Discussion

The experiments in Sect. 8.1 showed that centroid terms can be of benefit in both search and classification processes. While naturally centroid-based document retrieval cannot be used for finding exactly matching documents, more potential relevant documents can be found by this approach as the vocabulary mismatch problem is inherently solved. As an example, a search system applying it could return documents on cars, automobiles and vehicles when the query's centroid is 'car'. Due to its working principle, this solution is particularly useful in the fields of patent search and classification where besides exact document matches rough ones are of interest, too. A search system like DocAnalyser [58] whose task it is to find more similar and related web documents could benefit from this approach as well.

However, as indicated in Sect. 8.2, the most benefit will be gained when applying it in decentralised search systems, particularly when it comes to forwarding incoming queries of different diversity to peers with matching contents. The peers can make a routing decision based on the distance of those queries' centroids to the centroids of other peers (their contents is analysed for this purpose) while referring to the own (dynamically growing) reference co-occurrence graph which acts as a local knowledge base. Based on the results of the experiments in Sect. 8.1.5, this routing decision can be positively influenced by taking into account the neighbouring terms of the chosen centroid term in this graph as well the next best centroid candidates. These terms can be regarded as topical pointers as they provide a more precise semantic context and interpretation for the finally chosen centroid term, which might by itself not be the best representative for a query or document. By this means, an implicit word-sense disambiguation of the centroid terms is provided.

Furthermore, the peers can even use the technique of centroid-based document retrieval to answer incoming queries by themselves as the first set of experiments showed. However, for this purpose, other retrieval techniques based on the vector space model or probabilistic approaches might return results with a higher precision. Also, the application of highly performant and space-efficient Bloom filters [14] is a viable option at this search step. In order to gain benefit from all these retrieval approaches, their combined use might be sensible. In further experiments, this hypothesis will be investigated.

Last but not least, the experiments using English and German corpora presented in this book have shown that the methods of centroid-based text analysis and comparison are language-independent. A search system applying these approaches only needs to select the correct language-specific co-occurrence graph before the actual, language-independent query or document analysis is carried out.

8.5 Summary

This chapter investigated and evaluated several uses of text-representing centroid terms in search processes. It was found that the new method of centroid-based document retrieval is able to return both topically matching and relevant results. Also, an approach has been discussed to determine the correct word sense of a potentially ambiguous centroid term by enriching it with more specific terms.

In addition to these solutions, the two new graph-based measures diversity and speciality to qualitatively evaluate queries in interactive search sessions have been presented. These measures can instantaneously indicate whether a user should reformulate a query. Also, they are of value in subsequent processing steps such as the routing of queries in a tree-like document structure as obtained from the algorithms in the previous chapter.

Finally, a novel and fair method for user-based document ranking has been introduced and extensively evaluated. This method has been designed to be directly applied on such a document structure.

In the next chapter, the first P2P-based implementation of the 'Librarian of the Web', called the 'WebEngine', is introduced. The P2P-client's architecture, software components and core algorithms are discussed in detail.

Chapter 9
Implementation of the Librarian

9.1 Decentralised and Integrated Web Search

9.1.1 Preliminary Considerations

In the doctrine of most teachers and based on the users' experience, the today's WWW is considered a classical client/server system. Web servers offer contents to view or download using the HTTP protocol while every web browser is the respective client accessing content from any server. Clicking on a hyperlink in a web content usually means to be forwarded to the content, whose address is given in the URL of the link. This process is usually referred to as surfing the web.

Nevertheless, any web server may be regarded as a peer, which is connected to and therefore known by other peers (of this kind) through the addresses stored in the links of the hosted web pages. In such a manner, the WWW can be regarded as a Peer-to-Peer (P2P) system (with quite slow dynamics with respect to the addition or removal of peers). However, this system only allows users to surf from web document to web document by following links. Also, these restricted peers lack client functionalities (e.g. communication protocols) offered by web browsers such that there is usually no bidirectional communication between those peers possible (simple HTTP requests neglected which are mostly initiated by web browsers in the first place).

Moreover, as an integrated search functionality in the WWW is missing so far (the aforementioned restrictions might have contributed to this situation), centralised web search engines have been devised and developed with all their many shortcomings discussed in Chap. 3. Foremost, their approach of having to continuously copy and transfer web contents to their data centres causes avoidable network and processor load and is not power-saving. This problem will be inherently addressed by the subsequent implementation concept.

M. Kubek, *Concepts and Methods for a Librarian of the Web*,
Studies in Big Data 62, https://doi.org/10.1007/978-3-030-23136-1_9

9.1.2 Implementation Concept and WebEngine Prototype

In order to technically realise the new concept of the 'Librarian of the Web' in form of the mentioned decentralised, librarian-inspired web search engine, common web servers shall be significantly extended with the needed components for automatic text processing (clustering and classification of web documents and queries), for the processes of indexing and searching of web documents and for the library management. A general architecture of this concept can be seen in Fig. 9.1.

The concept scheme shows that a P2P-component is attached to standard web server. Its peer neighbourhood is induced by the incoming and outgoing links of local web documents. By this means, a new, fully integrated and decentralised web search engine is created.

In the following considerations, the name 'WebEngine' refers to the P2P-client software (P2P-plugin) and the P2P-network constructed by the respective clients.

As a prototype for this concept, the Java-based P2P-plug-in 'WebEngine' for the popular Apache Tomcat (http://tomcat.apache.org/) servlet container and web server with a graphical user interface (GUI) for any standard web browser has been developed. Due to its integration with the web server, it uses the same runtime environment and may access the offered web pages and databases of the server with all related meta-information. The following key points are addressed:

1. A connected, unstructured P2P-system is set up. Initially, the links contained in the locally hosted web pages of the Apache Tomcat server are used for this

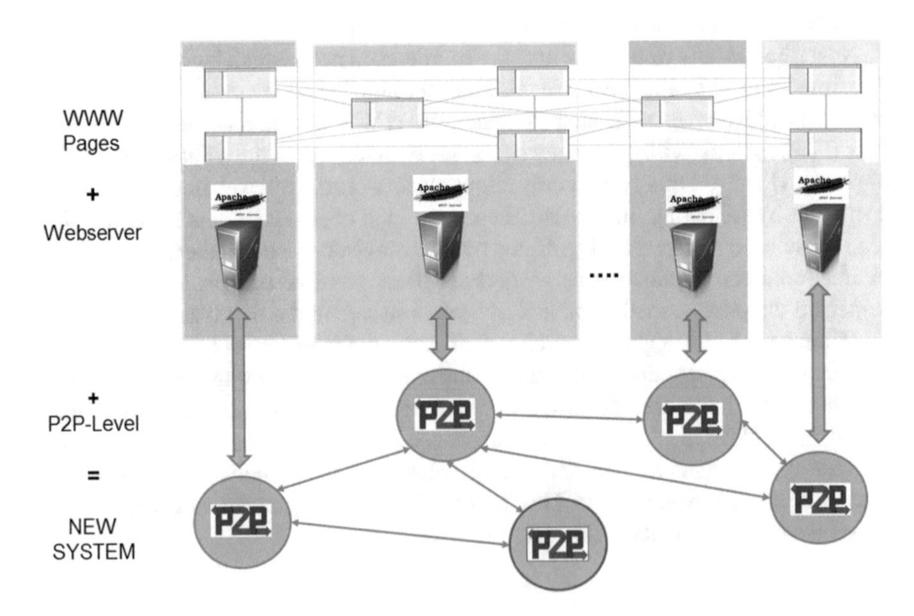

Fig. 9.1 First concept of a decentralised, integrated web search

purpose. Other bootstrap mechanisms as known from [57] and the *PING/PONG*-protocol from *Gnutella* and other P2P-systems may be applied at a later time, too. Note, that

- HTTP (HTTPS if possible) is used as frame protocol for any communication between the peers.
- A fixed number of connections between the peers will be kept open (although more contactable neighbours are locally stored).
- Furthermore, the mechanisms mentioned in Sect. 3.7.1 to limit the number of forwarded messages will be applied.

2. All hosted web documents will be indexed in separate index files after applied stopword removal and stemming.[1] The index is updated after every change in one of the hosted web documents. It acts as a cache to answer incoming queries in a fast manner. However, it would be sufficient—as shown in the previous chapter—to only store the centroid terms of local documents and their direct neighbours (their topical environment) in order to be able to answer queries properly.
3. The plug-in is able to generate a graphical user interface, in particular, a suitable search page for the requesting user (see Fig. 9.2).
4. Search results will be generated through a search in the local index files. Queries will also be sent via flooding to all opened connections to neighbouring peers. Here, the same methods to limit the number of forwarded messages as described in Sect. 3.7.1 are applied. Responding peers will return their results directly to the originating peer. Multi-keyword search is possible as well.
5. Proliferation mechanisms (as described in the next section) in the plug-in are integrated to support the distribution of the WebEngine-software over the entire WWW. The P2P-client is able to recognise the peer software on other web servers addressed and offer the download of its own program, in case the peer is not running at the destination yet.

The author hopes that the specified system rapidly changes the way of how documents are accessed, searched for and used in the WWW. The P2P-network may slowly grow besides the current WWW structures and make even use of centralised search engines when needed but may make them more and more obsolete. In this manner, the manipulation of search results through commercial influences will be greatly reduced.

9.1.3 Distributing the WebEngine

Nevertheless, for this purpose, it is necessary to reach a critical mass of peers partaking in this induced P2P-web. In order to reach this goal, the P2P-plug-in along with its

[1] In the first version of the P2P-plug-in, indexing is limited to nouns and names as the carriers of meaning.

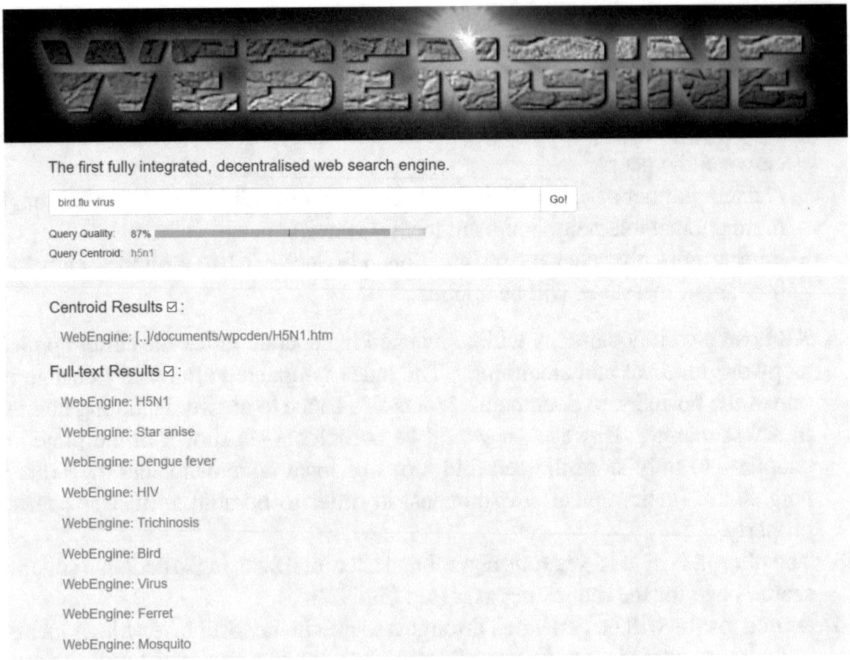

Fig. 9.2 The graphical user interface of the WebEngine

comprehensive installation and configuration instructions will be offered on a public website, from which users and especially administrators can directly download them. The P2P-plug-in itself will be provided in form of a .war (Web Application Resource) file, which can be easily deployed in the Apache Tomcat servlet container.

The download website will also offer an online service to instantly test the P2P-system without having to install any software, too. Also, the user interface of the P2P-plug-in will provide links to the software and social networks with options to share it. This way, any interested user coming across a web server running the P2P-software will be able to get to know more about it and will hopefully recommend it. Additionally, the WebEngine's search functionalities will be accessible using search fields in web pages such as weblogs. This way, a seamless integration is reached and users can instantly benefit from its services by just making use of the query input fields in the usual manner. As more and more people will recognise the mentioned new search functionalities on many different web sites, they will be made aware of the software offering them, too.

Another way to distribute the P2P-plug-in is getting in contact with web administrators not utilising the software yet. This can be achieved by monitoring the stream of visitors coming from websites. The P2P-software can e.g. analyse the HTTP headers of incoming requests stored in log files. This way, it can find out using the Referer-field the address of the previously visited web page linking to the current server. It can then automatically issue standard P2P protocol messages to the originating server in order to test if it is running the WebEngine already. If this is the case and if there is a significant amount of users coming from this server, it should store this server's address in the list of peer neighbours as this constant stream of users indicates a valid interest in both web server's contents.

If the originating server is not yet running the P2P-software, the current web server's P2P-plug-in can try to identify the respective web administrator using DNS WHOIS lookups or by analysing the imprint of the remote website in order to determine the name, e-mail address and postal address of its responsible webmaster using techniques known from natural language processing such as named-entity recognition. Once identified, the plug-in can suggest the current web server's administrator to get in contact with the other webmaster to introduce the P2P-software; for this purpose, the plug-in can conveniently provide an already filled contact form addressed to this person.

The same procedure can of course be applied for web servers that are frequently linked to by the current web server, too. Many outgoing links to a destination express its relevance to the local content. Therefore and especially in this case, it actually makes sense to contact the respective webmasters. Using these and further mechanisms, the P2P-plug-in can be distributed with little effort and the P2P-network will be established in a sustainable way by this means.

9.1.4 The Software Components of the WebEngine

At the beginning of this chapter, the general architectural concept of the WebEngine has been outlined and depicted in Fig. 9.1. Figure 9.3 shows the software components of the WebEngine-client. The blocks in the upper half of the scheme depict the functionality of currently running WebEngine prototype (basic implementation) presented in Sect. 9.1.2 with a particular storage facility to maintain the addresses of neighbouring peers.

The *Search Unit* is responsible to index local documents as well as to locally answer, forward and handle search requests issued by users. As mentioned above, in the WebEngine prototype, queries will be sent via flooding to all opened connections to neighbouring peers. However, a replacement of this basic procedure by a single-message, non-broadcasting, universal search protocol (USP), which forwards the search requests based on the centroid distance measure to the target node(s), is currently being integrated.

Analogously, in its lower half, Fig. 9.3 depicts the components which are yet to be integrated and tested at the time of writing this book (extension). This particularly concerns the components to carry out the centroid-based algorithms and derivations from them presented in the Chaps. 6 and 7 to construct, manage and maintain the hierarchical library structures.

The following components are currently being integrated:

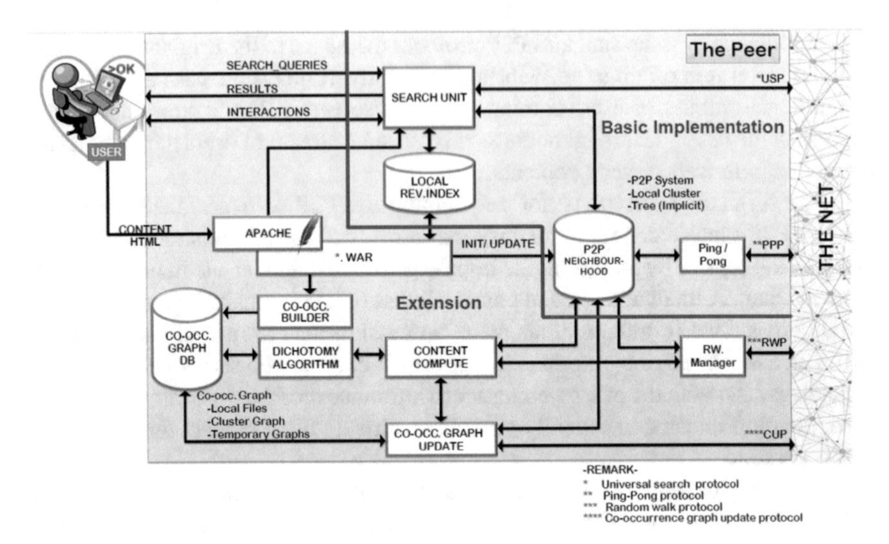

Fig. 9.3 The WebEngine's internal structure

- As the decentralised library management is—in contrast to the top-down algorithm presented in Sect. 7.2.1—carried out using random walkers, a particularly structured data unit circulating in the P2P-network, in the actual implementation of the WebEngine, a special *RW-Management* unit is added that carries out a special random walker protocol (RWP). Its working principles and methods will be described in detail in the next two sections. Also, the mentioned USP will be additionally extended such that random walkers will not only be used to generate the tree-like library but to perform search operations in it as well. For this purpose, special random walkers with query data as their payload will be sent out. This payload is matched and exchanged with other random walkers in the network when appropriate until matching documents are found. In this book, however, the decentralised library construction will be focussed, only.
- The processing of random walker data is performed by the *Content Compute* unit, which needs to access the co-occurrence graph databases (e.g. constructed and stored using the graph database system Neo4j (https://neo4j.com/)), which in turn contain the co-occurrence graph data of

  1. each web document offered by the local WWW server,
  2. the local term cluster which the node is responsible for,
  3. temporary operations of the random walker to build or update the hierarchic library structures.

- The remaining units support operations on and applying co-occurrence graphs:

  – In order to construct co-occurrence graphs, a co-occurrence graph builder is implemented in a separate unit.

– The needed clustering (see Sect. 7.2.2) is carried out by the *DichotomyAlgorithm* unit.
– the exchange of co-occurrence information is supported by the co-occurrence update protocol (CUP) controlled by a respective separate unit.

The next two sections extensively cover the concepts and algorithms for the decentralised library management using random walkers and explain, how graph databases are applied to support the data-intensive and content-related algorithms of the WebEngine.

9.2 Concept for Decentralised Library Management

The immediate application of the hierarchical clustering algorithm presented before requires the election of an initial and suitable first node, since subsequent merging activities may be difficult. This problem can be overcome by a novel bottom-up construction of the document hierarchy, which will be described below, although its origins go back to a contribution in 2002 [121]. Using methods of self-organisation and self-healing, the developed approach may even significantly contribute to the fault tolerance of the solution as, on the downside, the previously used classical tree-like architecture is characterised by its weak connectivity, too. At the time of writing this book, this method is being implemented in the WebEngine.

9.2.1 Model-Based Overview

In order to create a hierarchy of nodes (the library) in a bottom-up way, it is necessary to form groups of leaf nodes of the tree first. As this model-based overview shall be kept as generic as possible, the term 'node' represents an arbitrary instance (such a peer or document alike) in the communication structures. In the following considerations, the term 'primary group' is used to refer to a particular instance of such a group. Such a group consists of:

- N determined member nodes following a membership rule and
- max. ln(N)+1 representatives (surrogates) which are responsible for the hierarchic or flat communication between the groups.

A primary group has a flexible, performance-oriented size. A fitness function is applied for a permanent membership evaluation/re-assignment. Therefore, it is possible to speak of a vibrant group as well. By means of communication, the nodes coordinate themselves and cooperate for a dedicated goal (consensus to be reached), viz, the creation and management of groups with semantically close nodes (e.g. nodes that offer topically similar documents).

In order to actually form such a group of a predefined maximum size, it is sensible to send out special data units (tokens) by the nodes to their direct neighbours. A token contains information on the owner node (that sent out the token in the first place), a list of IP addresses of nodes in the group and content-related data such as the centroid terms of its nodes. The size of the group implicitly becomes apparent this way.

A receiving node first checks, if the maximum group size is not yet reached. If so, it can compare its topical profile with the one presented in the token of previously visited nodes. If it is semantically close, the node adds its own IP address and its content-related data to the token and in turn passes it to the next node referred to in the token or sends it back to the owner node. In order to be able to explore unvisited nodes, the token can be allowed to be sent to a certain number of randomly chosen neighbouring nodes. If after some circulations (intragroup communication) the token's data is not changed anymore, a stable group is created whose members can be found out by extracting this data. Figure 9.4 shows that a primary group formed this way is completely connected. In order to establish intergroup communication, the max. ln(N)+1 representatives (surrogates) can open communication channels (possibly long-distance links) with other clusters through their group representatives. Communication with other members in these groups is possible via their respective surrogate node(s) only.

This flat interconnection is depicted in Fig. 9.5. Also, small-world communication structures will appear this way. However, as it is necessary to form tree-like structures, a hierarchical interconnection must be established. For this purpose, a primary group is represented by a new instance of the surrogate node which is virtually dou-

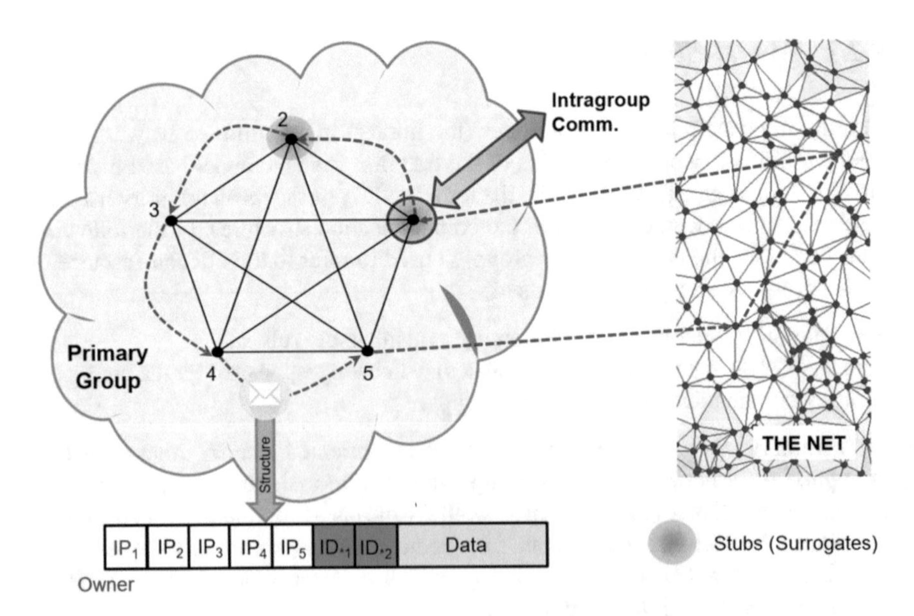

Fig. 9.4 Forming a primary group

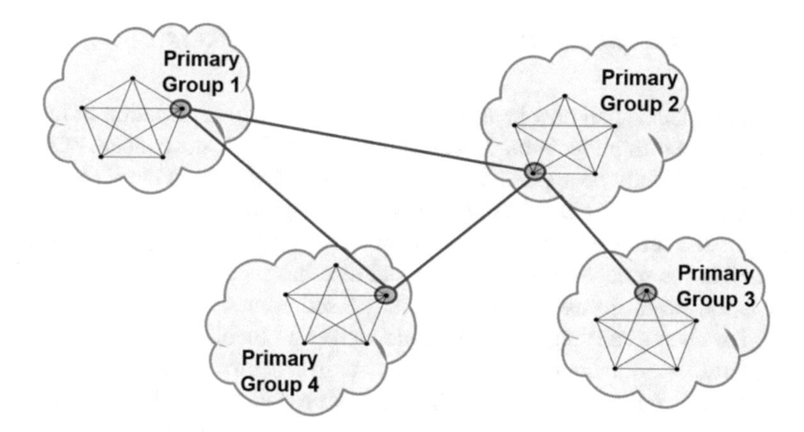

Fig. 9.5 Connecting primary groups

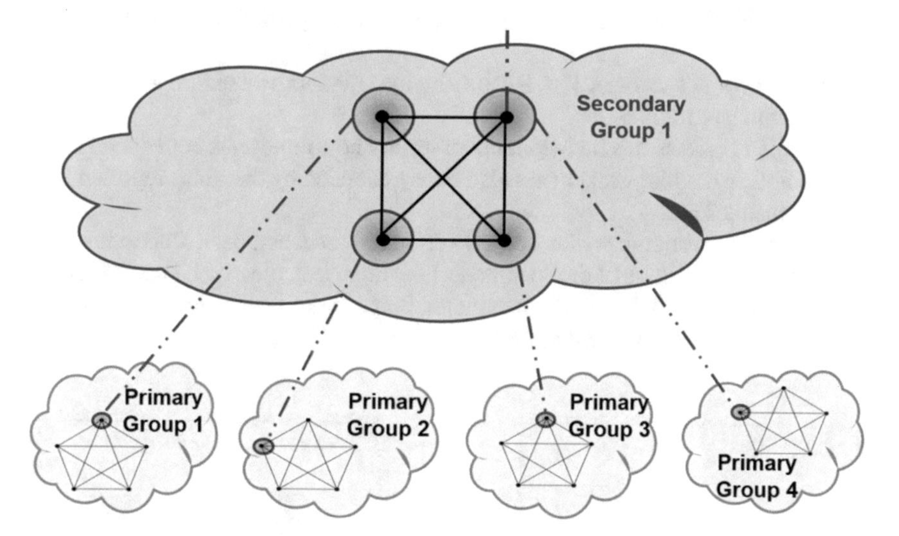

Fig. 9.6 Creating group hierarchies

bled when the group is stable for a certain amount of time. Both node instances may communicate with each other. More importantly, the new instance defines a new, independent communication level (in this case the second). Furthermore, communication is possible between surrogate nodes on the same level only.

As Fig. 9.6 shows, a secondary group is formed by these doubled surrogate nodes. In such a group, tokens are passed around by the nodes on this communication level only. When a secondary group is found to be stable, its surrogate node is doubled again for the next-level group building. By this means, the needed tree-like communication structures appear.

9.2.2 Library Construction Using Random Walkers

This generic model shall now be used as a basis for a vivid explanation of the algorithms in the next three sections. As indicated in Sect. 9.1.4, the concept of random walkers shall actually be applied by the *RW-Management* unit in the WebEngine to create and maintain the tree-like library in a bottom-up manner. A random walker is a specially structured data unit circulating in the P2P-network and acts as the token mentioned in the model-based overview and also has its data fields. Each random walker is, however, no executable program passed around but is managed by the WebEngine-clients carrying out the random walker protocol (RWP) instead.

Generally, the following steps are carried out in order to create this library.

At first, for each document in the network, a random walker is sent out. If two random walkers coincidentally meet at any peer, they can be merged (while cancelling one of them in the process) if the sum of the number of the filled positions in both of their lists of IP addresses is smaller or equal than the predefined maximum group size.

This process is shown in Fig. 9.7 for the two random walkers representing the initial documents 1 and 3.

As seen in Fig. 9.8, this merged random walker now represents both documents 1 and 3. Another random walker (has also been generated by merging) represents the two documents 2 and 4.

These merged random walkers could—in turn—meet at a peer. Due to the computed semantic closeness of the documents they represent, they exchange the respective links (here, the links to the documents 2 and 3 are exchanged) such that the content similarity is maximised (see Fig. 9.9).

Fig. 9.7 Generating and merging random walkers

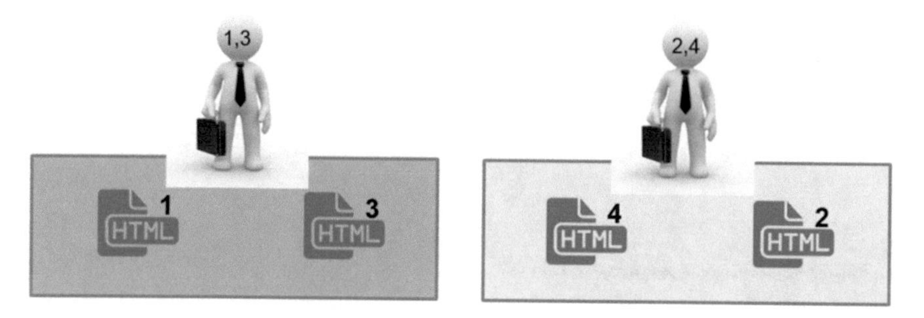

Fig. 9.8 Sets of documents represented by merged random walkers

Fig. 9.9 Merged random walkers exchange their contents

Fig. 9.10 Semantically closer document groups managed by the random walkers

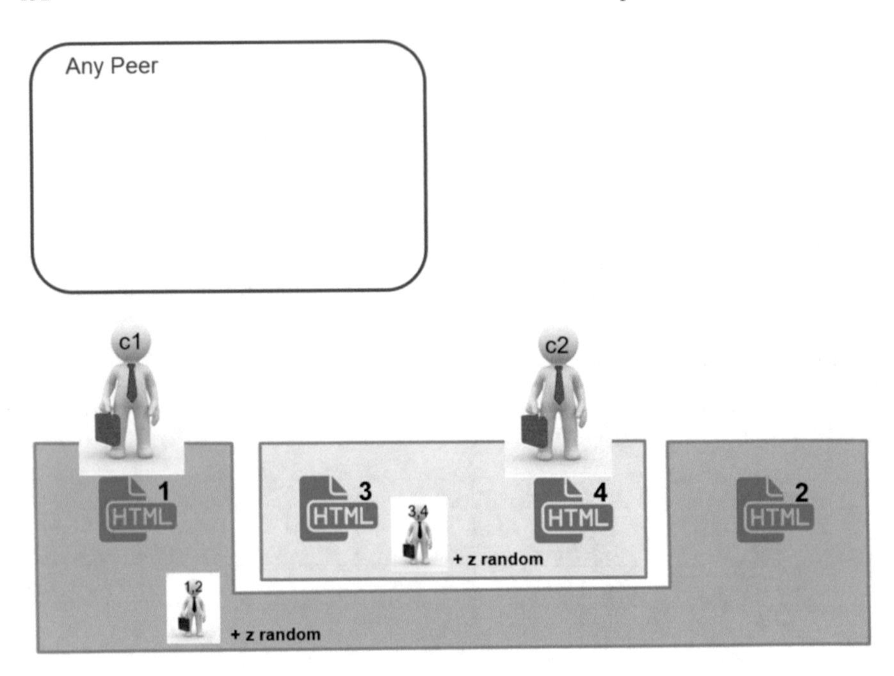

Fig. 9.11 Generating random walkers at the next higher level

As depicted in Fig. 9.10, the random walkers represent semantically closer groups of documents afterwards. These document sets form the previously mentioned primary groups (level 1).

After some time, it is detected by continuously inspecting the random walker data that these groups remain stable (see Fig. 9.11). Therefore, new random walkers at the next higher level 2 representing them are generated.

If these random walkers meet at a peer (Fig. 9.12), they might be merged in the same way as described for the random walkers at level 1. However, this process is only carried out for random walkers at the same level.

As Fig. 9.13 shows, this newly merged random walker at level 2 is now responsible for the complete document set in this example. However, the random walkers at level 1 keep circulating in the network while they are allowed to be sent to a certain number of z randomly chosen neighbouring peers (z positions in the random walker are reserved for this purpose) in order to keep track of possibly occurring content changes in the P2P-network.

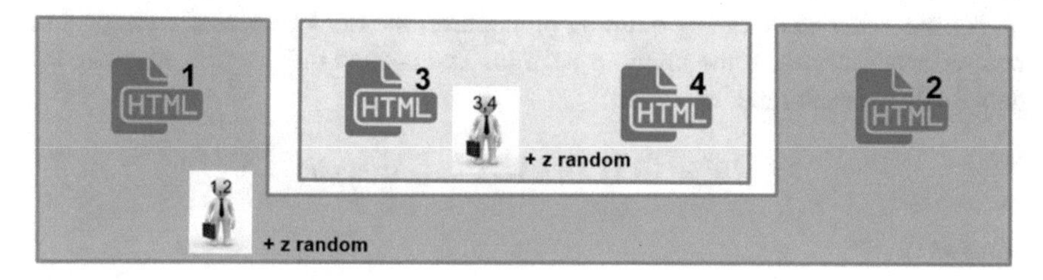

Fig. 9.12 Meeting of higher-level random walkers

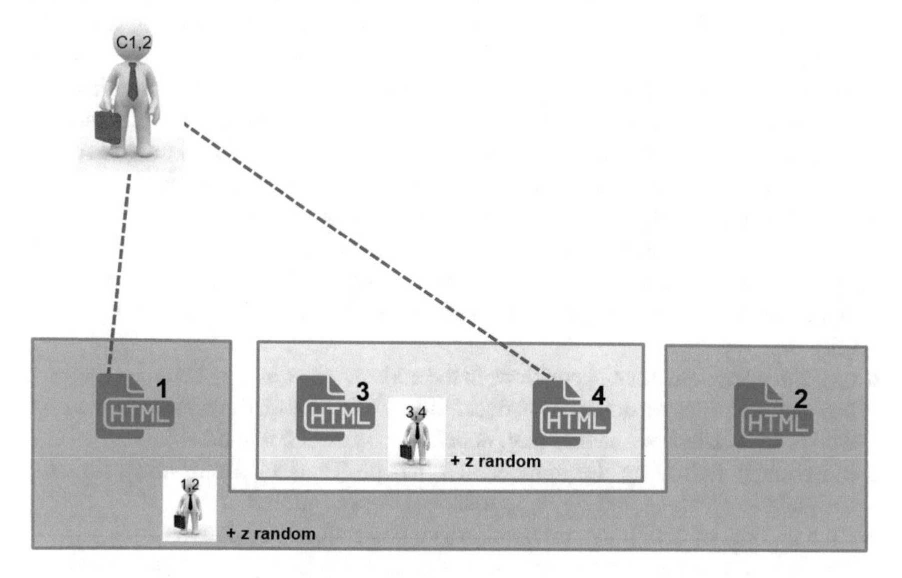

Fig. 9.13 Merged, higher-level random walker

9.3 Algorithms for Decentralised Library Management

9.3.1 Handling of Random Walkers

Now, the actual P2P-system is considered, which is built and formed by the web server extension WebEngine as described before and in [31]. Every peer P with the IP address $IP(P)$ of this system has its own neighbourhood warehouse $N(P)$ containing initially the neighbours derived from the links of pages hosted by the web server. It is updated by the known, standard P2P ping-pong-protocol.

For the intended hierarchy building procedures, the random walker concept shall be used as described in the previous section. The needed information structure (all data fields) of each random walker

$$RW = (IP[], D[], W[], c_{max}, c, z, ptr)$$

consists of

- an array $[IP_1, IP_2, \ldots, IP_{c_{max}}]$ of positions to store IP addresses of peers belonging to a local cluster on a given level; whereby IP_1 always denotes and is initialised with the IP-address $IP(P)$ of the owning peer, i.e. the peer which has generated the random walker,
- a second array $[D_1, D_2, \ldots, D_{c_{max}}]$ which contains the respective, remaining path information (locator) to identify the document on the corresponding peer (needed if a peer offers more than one document). Position D_1 is again reserved for the locator of the document of the owner/generator of the random walker.
- a third array $[W_1, W_2, \ldots, W_{c_{max}}]$ containing an ASCII content description of the documents represented by the locator in the corresponding positions of $IP[] : D[]$ (and subsequent documents or sub-trees) managed by this random walker.[2] In the course of this elaboration, the centroid term of the document will be used for this purpose.
- c_{max} the overall number of positions in the address array field which may be either a global constant or determined depending on the network conditions for each walker in an adaptive manner (e.g. depending on circulation times).
- c a counter indicating the number of filled positions in the IP array (initially assigned to 1). Note, that $c \leq c_{max}$ must be always fulfilled.
- z the number of additional, randomly to be determined positions.
- ptr the index of the current position of the walker with $1 \geq ptr \geq (c_{max} + z)$ and initialised to 1.

In the beginning, one random walker is generated for each document available from the respective web server by the corresponding peer. Every document will be made identifiable besides $IP(P)$ by a local document locator D_h. An empty set $X(D_h)$

[2]Note, the random walker may be more light-weighted, if this information is not a part of the random walker but retrieved from the peer of origin.

is stored, which later will contain IP addresses of peers with similar contents, i.e. which will form a completely connected sub-cluster.

Furthermore, each peer belonging to a web server executes the following algorithm to process the generated and/or arriving random walkers:

1. $LOOP_1$:
   Receive a new random walker $R_n = recv()$.
2. Determine randomly the earliest departure time $\tau(R_n)$ for the random walker.
3. Use a set C to keep all pairs of random walkers at the computer and let $C = \emptyset$ at the start time of the algorithm.
4. Let R be the set of random walkers, recently visiting the considered peer. Initialise $R := R_n$.
5. $LOOP_2$:
   Check $\forall i := 1 \ldots |R|$:
   IF $\forall j := 1 \ldots |R| \wedge (i \neq j)$

   - $(R_i, R_j) \notin C$ and
   - $(R_j, R_i) \notin C$ and
   - the recent time $t > \tau(R_i)$

   THEN

   - Set $R := R \setminus \{R_i\}$.
   - If $(ptr \leq c)$ then set $X(D_{ptr}) = \{IP_1, IP_2, \ldots, IP_c\}$.
   - Increase $ptr := ((ptr + 1) mod (c_{max} + z)) + 1$.
   - Send out R_i to IP_{ptr} by $send(IP_{ptr}, R_i)$, if $ptr \leq c$ and otherwise to a randomly chosen successor out of the set of location in the neighbourhood set of the peer $IP_{next} \in N$ by $send(IP_{next}, R_i)$, if $c < prt \leq (c_{max} + z)$.

6. If another random walker R_s is received

   a. Determine randomly the earliest departure time $\tau(R_s)$ for this random walker.
   b. Set $C := C \cup (\{R_s\} \times R)$.
   c. Update $R := R \cup \{R_s\}$.

7. If $C \neq \emptyset$

   a. Take any $(R_i, R_j) \in C$.
   b. $Compute(R_i, R_j)$.
   c. Set $C := C \setminus \{(R_i, R_j)\}$

8. If $R = \emptyset$ GoTo $LOOP_1$ otherwise GoTo $LOOP_2$.

In order to ensure fault tolerance, any document D_i which is not checked by its random walker within a given timeout period T_{out} may take activities to synchronise a new random walker using a standard election mechanism with all nodes within $X(D_i)$.

9.3.2 Computation of Random Walker Data

This section describes the needed computation $Compute(R_i, R_j)$ for any pair of random walkers, which are meeting at any node in the P2P-system. To distinguish both data areas, they are denoted by a leading first index in the formulae.

For the computation, two cases can be distinguished.

1. $c_i + c_j \leq c_{max}$
   i.e. the two random walkers may be merged into a single one. This case mostly appears at the start of the system or if a set of new documents and/or peers is added.
   The following updates are carried out for merging:
   a. $\forall k = (c_i + 1) \ldots (c_i + c_j) : IP_{i,k} = IP_{j,k-c_j}$
   b. $\forall k = (c_i + 1) \ldots (c_i + c_j) : D_{i,k} = D_{j,k-c_j}$
   c. $c_i = c_i + c_j$
   d. $c_{max,i}$, ptr_i and z_i remain unchanged
   e. The random walker R_j is cancelled by
      - $\forall (j := 1 \ldots |R|) \wedge (i \neq j)$
        let $C := C \setminus (R_i, R_j)$ and
        let $C := C \setminus (R_j, R_i)$.
      - $R := R \setminus \{R_j\}$

2. $c_i + c_j > c_{max}$
   i.e. the two random walkers met cannot be merged but the attached document links shall be sorted such that a maximum similarity is reached. Therefore the documents addressed via the set of URL's $(IP_n, m : D_n, m)$

$$U = \bigcup_{m=1}^{c_i} \{(IP_{i,m} : D_{i,m})\} \cup \bigcup_{m=1}^{c_j} \{(IP_{j,m} : D_{j,m})\}$$

   are considered. Any clustering method or dichotomy building algorithm using document centroids in $W[]_i$ and $W[]_j$ as described before in Chap. 7 shall be used to generate two subsets U_i and U_j from U with $|U_i| < c_{max,i}$ and $|U_j| < c_{max,j}$. The co-occurrence graph used for this purpose can be either obtained from the existing sub-cluster i or j or be a (temporary) combination of both.
   The following updates are carried out for merging R_i and R_j using U_i and U_j:
   a. $\forall k = 1 \ldots |U_i| : IP_{i,k} = IP_{i,k}(U_i)$
   b. $\forall k = 1 \ldots |U_j| : IP_{j,k} = IP_{j,k}(U_j)$
   c. $\forall k = 1 \ldots |U_i| : D_{i,k} = D_{i,k}(U_i)$
   d. $\forall k = 1 \ldots |U_j| : D_{j,k} = D_{j,k}(U_j)$
   e. $c_{max,i}$ and $c_{max,j}$ remain unchanged.
   f. Set $c_i = |U_i|$ and $c_j = |U_j|$
   g. z_i and z_j remain unchanged.

h. $ptr_i = c_i + z_i$ and $ptr_j = c_j + z_j$ to ensure that the updates are executed in the fastest manner starting at the first node of the cycle in the next step.

In both cases, the notification of all participating nodes on the performed changes is sent out. Also, an update of the local clusters is executed within the next circulation of the random walker.

In addition, the co-occurrence graphs of all nodes related to the newly merged random walker are merged as well and the respective resulting centroid terms of all documents are determined under the management of the peer owning the random walker (i.e. which is on $IP[1]$). Since this procedure as well as the future use of the owner (first peer in the sequence) of a random walker requires some computational performance of the respective machine, changes in the order of $IP[]$, $D[]$ as well as $W[]$ might be indicated and useful.

9.3.3 Building the Hierarchy

9.3.3.1 Structural Hierarchies

The methodology introduced in the previous sections generate completely connected clusters of the most similar documents, which are still isolated substructures. Now, they shall be connected in a tree-like, hierarchical structure of completely connected sub-graphs.

Therefore, an addition must be made to the definition of random walkers

$$RW = (IP[], D[], W[], c_{max}, c, z, ptr)$$

by changing it to

$$RW = (IP[], D[], W[], c_{max}, c, z, ptr, lev)$$

whereby lev, represents the level of the tree, on which the respective random walker is acting. Note that the leaves will have—differing from usual enumerations—level 1.

For the further processing steps, the following updates are necessary:

1. At the starting point, of the above described algorithm, $lev = 1$ represents the lowest level (the document level).
2. In W, the centroid term of the associated document is indicated, when random walkers are merged. It is replaced by the centroid term of all documents represented by the random walker.
3. The owner of the random walker, i.e. usually the peer represented in the first position of $IP[]$, observes all changes in the random walker. If for a fixed, longer time interval Δ

- no changes are observed in the $IP[]-$ and $D[]-$area of the random walker and
- $1 < c \leq c_{max}$,

the $IP[1]$ peer is allowed *to launch another, new random walker.*

4. By doing so, it may happen that on any level a single random walker may not be connected to the community when all existing other random walkers have filled exactly all c_{max} positions.
To avoid this case, position $c_{max} + 1$ is used as a temporary (emergency) position. If $c = c_{max}$, it may merge with a random walker with $c = 1$ but must release this position at the next possible solution regardless of any content-related aspects, i.e. as soon as either a random walker with $c < c_{max}$ is met or another random walker with $c_{max} + 1$ filled positions is found (in this case a new random walker at the same level is created).
5. This *new random walker* follows exactly the rules for the initial settings, as described above for the level 1-random walker, with the following updates:

   - The respective level information is derived from the level information of the generating walker and increased by 1.
   - W contains the centroid term which is calculated from the centroid terms of all participating (represented) documents (level 2) or random walkers (in the higher levels), which are usually stored on the generating peers.

6. Also, the behaviour of the new random walker follows exactly the rules described above. However, a computation, merging and sorting will only take place for random walkers having the same level lev.
7. For reasons of fault tolerance, the $IP[]-$ and $D[]-$area may contain instead of a single information a sub-array, containing a small number (i.e. two or three) of information from other nodes of the represented sub-cluster. In such a manner, an unavailability of the first peer in the list may be tolerated by using a replacement peer.
8. The described hierarchy building mechanism is successively repeated for all higher levels 2, 3, ... and stops automatically, if the conditions formulated in 3 cannot be fulfilled anymore.

After these steps have been carried out, the needed tree-like library is obtained. Its structure is stored in data fields of the random walkers circulating in the P2P-network.

9.3.3.2 Hierarchies of Co-occurrence Graphs

Any decision making is based on co-occurrence graphs stored locally. On the lowest level 1, the respective co-occurrence graphs are built from one document. As soon as the positions $1 \ldots c_{max}$ of each random walker are filled, temporary and bigger co-occurrence graphs may (will) be built and used for decision making but must eventually be re-organised, since every document may only represent its set of co-occurrences in one co-occurrence graph $R_{c,lev}$ on each level lev.

With the generation of a random walker for $lev + 1$, the co-occurrence graph of the respectively represented region on level lev shall be built and be stable. At least with the complete compositions of the random walkers on level $lev + 1$, also the next level co-occurrence graph $Rc, lev + 1$ shall be available and will be used in the same manner to assemble the next hierarchy level $lev + 2$. In addition, $Rc, lev + 1$ will be handed down to all random walkers until level 1.

Of course, the respective (but seldom necessary) updates may cause repeated evaluations of the similarity of documents (sets of documents in different regions) combined in a random walker. Since the remaining z randomly chosen positions in each random walker will allow meetings at any peer at any time, these discrepancies will automatically be recognised and the structure will be adapted. Exactly the same process will result in an automatic inclusion of newly appearing documents and their random walkers.

Note that any change in a random walker requires a new calculation of the co-occurrence graph on the respective level with the corresponding re-calculations of its upper and lower levels. However, this process is relatively seldom needed (last but not least to the stability of co-occurrence graphs as previously pointed out in Sect. 6.4). Due to its size, all Rc, lev shall never be a part of a random walker but solely be kept in a graph database of the represented nodes, regions or sub-clusters. The complete connection of those structures (as described above) may make those updates even more simple.

9.4 Graph Databases—Handling Large Graphs Effectively and Efficiently

When taking a look at the WebEngine's architecture and functionalities from a technological point of view, it becomes obvious that it is necessary to be able to manage large graph structures efficiently and effectively. Graph database systems such as Neo4j (https://neo4j.com/) are specifically designed for this purpose. Especially, they are well-suited to support graph-based text mining algorithms [87]. This kind of databases is not only useful to solely store and query the herein discussed co-occurrence graphs, with the help of the property graph model of these databases, nodes (terms) in co-occurrence graphs can be enriched with additional attributes such as the names of the documents they occur in as well as the number of their occurrences, too. Likewise, the co-occurrence significances can be persistently saved as edge attributes. Graph databases are therefore an urgently necessary tool as a basis to realise the herein presented technical solutions and support all graph-based algorithms discussed.

Generally, graph databases will play an increasing role at the intersection of both large-scale local and global information retrieval and text mining in the future. As an example case, Google's so-called 'Knowledge Graph' [38] can be regarded as an entity-centric knowledge base and has been applied as a search-augmenting technology ever since its introduction. Its comprehensive set of modelled relations between a large variety of entities sustainably enriches the usual linear result lists whose drawbacks have been discussed in detail. Thus, it can be clearly seen that turning away from simple token-level information retrieval is of benefit for the users.

Also, as discussed in the Chaps. 4 and 5, many techniques for (graph-based) natural language processing and text mining have been proposed in the last years. However, it can be noted that their integration with technological advanced search technologies is still uncommon. Here, graph databases will play an important role at the technological level in the future. For instance, search systems relying on them will—depending on the data model used—not only be able to return matching results to users' queries but to instantly present semantic relationships e.g. as clusters between them as well. Thus, in order to provide actually useful search results, it is mandatory that future search and information systems apply text mining algorithms as well. Here, graph-based approaches are of particular interest.

This book presented several newly developed algorithms and applications that implement this integration on two levels: analysis and synthesis. On the one hand, the introduced algorithms for graph-based keyword extraction and text clustering have been designed to directly fulfil specific search tasks such as finding similar and related web documents. On the other hand, they have been designed and applied for the purpose of structure building and the semantically oriented routing of queries (see Chap. 7 in particular). In other words, semantic relationships affect the way, how search processes are carried out. Therefore, the WebEngine's implementation makes especially use of embedded Neo4j graph databases for the storage, traversal and clustering of co-occurrence graphs and web documents.

9.5 Further Considerations and Business Models

While each WebEngine's tasks are primarily aimed at maintaining and semantically structuring the locally provided web contents, at the same time and usually, a topically specialised knowledge base (the tree-like document structure) is implicitly constructed which also involves connected peers, the child nodes.

The access to these tree parts—one could speak of an artificial expert network, too—might be of particular help in long-term or in-depth (re)search processes. As already pointed out in this book, these processes are currently insufficiently supported by existing web search engines. Although it is intended by the author that the WebEngine's functions are provided free of charge, webmasters might out of various reasons be interested in restricting the access to certain parts of the tree-like structure and provide it to a number of selected peers or users, only. As an example, access rights can be granted when a certain amount of money has been paid or a valid registration is available.

Registered peers or users that wish to gain access (login) to a WebEngine's library and associated library subtrees for a search session have to provide valid credentials such as usernames, passwords or pre-shared keys. The validity of these credentials can be restricted (e.g. by granting access for a certain amount of time or by limiting the number of allowed requests to specific subtrees) as well. Furthermore, in order to encrypt (i.e. digitally sign) these credentials along with the returned answers as well as to ensure the authenticity of the senders and the integrity of the exchanged messages, a public key infrastructure (PKI) must be set up and respective functions (e.g. for the management and distribution of certificates) integrated into WebEngine. Along with these technical prerequisites, in future versions of the WebEngine, various options for the management of access rights and user accounts as well as functions for the registration, authentication and authorisation of users and peers will be implemented.

9.6 Summary

This chapter presented the P2P-based implementation of the 'Librarian of the Web', called WebEngine. Its features, local working principles and software components have been elaborated on in detail. The main advantage of the WebEngine is that it utilises existing web technologies such as web servers and links in web documents in order to create a fully decentralised web search system. As an extension to the well-known Apache Tomcat web server, it is easy to install and maintain for administrators. Furthermore, implementation-specific adaptations and extensions of the algorithms presented in Chap. 7 have been discussed. As the WebEngine relies on graph-based text analysis techniques applied on co-occurrence graphs, the graph database system Neo4j (https://neo4j.com/) has been used for the persistent storage and retrieval of terms, links to documents as well as their relations.

Chapter 10
Conclusion

10.1 Main Results

Recent search and information retrieval procedures in the today's World Wide Web have been analysed. Substantial disadvantages have been identified, which show that Google and other centralised search engines are not sufficient tools to master future challenges in this field. In particular, manifold services of old-fashioned libraries and tasks of librarians working in there have been identified and presented, whose realisation and application in the context of the web could drastically improve the mentioned situation.

It was clearly worked out that only a decentralised approach using P2P-technology will be a promising one. The developed concept has been divided in two major steps. First, significant contributions have been presented to sustainably support web search by locally working tools and services in cooperation with the current search engine concept. With FXResearcher and DocAnalyser, concepts and tools have been developed, which analyse (potentially sensitive) textual data of the user on the local computer or take into account currently viewed web documents, only. This information may significantly help to make keyword-based queries more precise by properly extending them or automatically generating queries of four terms to ensure an improved result quality. Also, the possibility to process user feedback was included in those methods. They allow for an interaction with the user during search sessions, which is now considered as a process with a determined history and specific search contexts. The result presentation was significantly enhanced using graphical methods, e.g. in the tool PDSearch. Last but not least, mobile devices were supported, too. With Android IR, an approach for power-saving and integrated full-text search on Android devices has been presented. Nevertheless, all these works were only evolutionary steps towards an improved search in the web.

A major progress has been made by the introduction of text-representing centroid terms. Although those terms can be clearly formally calculated, as a well-balanced extract in form of a single word or term they represent typical content aspects of (longer) documents as well as (short) search queries of arbitrary length. Some major

M. Kubek, *Concepts and Methods for a Librarian of the Web*,
Studies in Big Data 62, https://doi.org/10.1007/978-3-030-23136-1_10

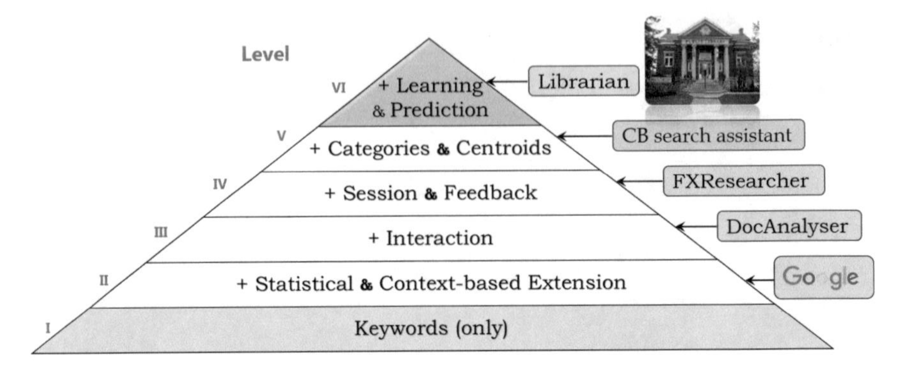

Fig. 10.1 Different levels of search

properties of such centroid terms have been worked out and were subsequently analysed. A method for the fast calculation of centroid terms has been developed using the spreading activation technique. Due to its local working approach, it reduces the processing time to less than a second even for large texts and co-occurrence graphs consisting of several thousand nodes. It was shown that centroid terms can be also used to detect content similarities even when—in contrast to outdated bag-of-words approaches—there is a vocabulary mismatch. Finally, the centroid distance measure based thereupon has been successfully applied to derive two centroid-based, hierarchical clustering methods. In comparison to typically used methods, their overall cluster quality increased by on average 35 percentage points.

Returning to the major work of a classic librarian to establish and structure a collection of books, it was shown that these innovative approaches can result in a completely new kind of semantically connecting, managing and searching documents in the web. The book explained that these are no longer separated tasks in the future but shall be carried out by a properly extended web server. The work documented the way from a classic client/server-based WWW to a real peer-to-peer system. Therefore, every web server employs a peer managing its own contents and connects it in the right manner with other peers. In a fully decentralised manner, a tree-like hierarchy of documents can be built bottom-up and managed by a population of random walkers. This way, a flexible and fault-tolerant P2P-system is created which also avoids the typically needed document crawling and copying processes known from centralised search engines. It has also been shown that even an adaptive, user access-based document ranking can be established in the context of the structures built this way. Finally, diversity and speciality have been identified as parameters, which allow for a qualitative evaluation of search terms and therefore may guide the user in the search process.

While a respective prototype P2P-client, called WebEngine, has been introduced, the complete implementation of all the described methods is in progress and will be completed in 2019.

The pyramid of search levels in Fig. 10.1 depicts the achieved advantages of the centroid-based (CB) search assistance in an impressive manner. Just a last step needs to be taken towards a real librarian of the web: the recently started development must be completed by adding learning and predicting approaches. Together with a continuous extension and refinement of the methods described in this book, these additions will constitute the major parts of future works.

10.2 Outlook

The introduced librarian-inspired approaches, methods and technical solutions present a major step towards a fully decentralised and integrated web search. Their aim is to actually support users in diversely oriented search sessions. Likewise, there are many ways and ideas to improve them:

1. Future works will deal with further improvements of the document clustering approaches and the decentralised structure-building methods to overcome problems such as increased query routing times caused by a potentially low node connectivity in the P2P-network. For this purpose, the list of IP adresses and content-related data fields of random walkers will be used to establish long- and short-distance links to other peers in the network. Also, random walkers will be applied to carry out search tasks as described in Sect. 9.1.4.
2. As the concepts of centroid terms and the centroid distance measure reside at the core of the herein presented text analysis and clustering solutions, future research will be focussed on even faster calculation methods for centroids while considering the human (sequential) reading process. Also, it will be investigated, if the derived technique of centroid-based document retrieval can be successfully used in conjunction with traditional information retrieval methods and web search approaches in order to get benefit from all their approaches. As with the solutions presented in Chap. 5, it is conceivable to combine the centroid-based techniques with classic web search engines such as Google to locally rerank, reorder and topically group returned search results in real-time. These modified results would then be more user-specific, too.
3. Novel user interfaces to actively and sustainably support interactive web search sessions also need to be developed which allow for a targeted navigation in the user's search history and in the currently presented search results. For this purpose, different visualisation approaches must be conceived and evaluated. Consequently, a search system such as the WebEngine must then be able to learn from implicit and explicit user feedback, i.e. observe and analyse the continuous interaction with its users or enter into a dialog with the user when appropriate. These approaches would present a major step forward to turn the system into an automatic signpost guiding users to relevant information in the web.
4. In this context and in continuation to the approaches presented in this book, it is advisable to think about new solutions and flexible models for the management of

graph data, especially terms and their relations. As an example, for the purpose of steadily making relevant predictions about the next search steps and good routing decisions as discussed before, these solutions have to be responsible for the detection of new as well as the strengthening and weakening (modification) of existing term and topic relations. Also, functions for their persistent storage and retrieval need to be added accordingly. In this regard, it is sensible to determine if and when to intentionally delete respective relations entirely, e.g. to model digital obsolescence or oblivion.

5. The addition of special-purpose search functions along with the integration of collaborative features is another way to enrich those systems. As an example, extracted source topics of documents provide the possibility to track topics to their roots by repeatedly using them as search words. A dedicated function for this purpose would make research in the (digital) humanities much easier. Also, means to collaboratively (e.g. in expert groups) search for or investigate topics of interest are of high interest for specific user groups.

As it is easy to conclude from the presented scientific results in this book and the aforementioned improvement suggestions, the 'Librarian of the Web' is a long-term research and development project which will be steadily cared for by the author and will hopefully inspire research scientists and developers around the world to contribute to its success.

References

1. Agirre, E., et al.: SemEval-2016 Task 1: Semantic Textual Similarity, Monolingual and Cross-Lingual Evaluation. Proceedings of SemEval-2016 (2016)
2. Agrawal, R., et al.: Enrichment and reductionism: two approaches for web query classification. In: Proceedings of International Conference on Neural Information Processing (ICONIP 2011), Volume 7064 of Lecture Notes in Computer Science, pp. 148–157. Springer, Berlin, Heidelberg (2011)
3. American Library Association: Presidential committee on information literacy: Final report. http://www.ala.org/acrl/publications/whitepapers/presidential (1989). Accessed Feb 10 2018
4. Al-Sharji, S., Beer, M., Uruchurtu, E.: A dwell time-based technique for personalised ranking model. In: Chen, Q., Hameurlain, A., Toumani, F., Wagner, R., Decker, H. (eds.) Database and Expert Systems Applications. Lecture Notes in Computer Science, vol. 9262. Springer, Cham (2015)
5. American Library Association: Guidelines for Behavioral Performance of Reference and Information Service Providers. http://www.ala.org/rusa/resources/guidelines/guidelinesbehavioral (2013). Accessed Jan 5 2018
6. Anderson, C.: The Long Tail: Why the Future of Business Is Selling Less of More. Disney Hyperion (2006)
7. Baeza-Yates, R., et al.: The impact of caching on search engines. In: Proceedings of the 30th Annual International ACM SIGIR Conference on Research and Development in Information Retrieval (SIGIR '07), pp. 183–190. ACM New York, NY, USA (2007)
8. Baeza-Yates, R., Ribeiro-Neto, B.: Modern Information Retrieval. 2nd Edition, Addison-Wesley Publishing Company (2011)
9. Bell, J.: The developing role of librarians in a digital age. http://www.infotoday.eu/Articles/Editorial/Featured-Articles/The-developing-role-of-librarians-in-a-digital-age-110185.aspx (2016). Accessed Jan 5 2018
10. Berners-Lee, T., et al.: The semantic web. Scientific American **284**(5), 34–43 (2001)
11. Biemann, C.: Chinese whispers: an efficient graph clustering algorithm and its application to natural language processing problems. In: Proceedings of the HLT-NAACL-06 Workshop on Textgraphs-06, pp. 73–80. ACL, New York City (2006)
12. Biemann, C., Bordag, S., Quasthoff, U.: Automatic acquisition of paradigmatic relations using iterated co-occurrences. In: Proceedings of the 4th International Conference on Language Resources and Evaluation (LREC 2004), pp. 967–970. Lisbon, Portugal (2004)
13. Blei, D.M., Ng, A.Y., Jordan, M.I.: Latent Dirichlet allocation. J. Mach. Learn. Res. **3**, 993–1022 (2003)
14. Bloom, B.H.: Space/time trade-offs in hash coding with allowable errors. Commun. ACM **13**(7), 422–426. ACM New York, NY, USA (1970)
15. Brisbane, A.: Speakers give sound advice. In: Syracuse Post Standard (March 28 1911, p. 18) (1911)

M. Kubek, *Concepts and Methods for a Librarian of the Web*,
Studies in Big Data 62, https://doi.org/10.1007/978-3-030-23136-1

16. Broder, A., et al.: Graph structure in the web: experiments and models. In: Proceedings of the 9th International World Wide Web Conference on Computer Networks: The International Journal of Computer and Telecommunications Networking, pp 309–320. Amsterdam, The Netherlands (2000)
17. Budanitsky, A., Hirst, G.: Evaluating wordnet-based measures of lexical semantic relatedness. Comput. Linguist. **32**(1), 13–47 (2006)
18. Büchler, M.: Flexibles Berechnen von Kookkurrenzen auf strukturierten und unstrukturierten Daten. Master's thesis. University of Leipzig (2006)
19. Capoccia, A., et al.: Detecting communities in large networks. Physica A: Stat. Mech. Appl. **352**(2–4), 669–676. Elsevier, Amsterdam (2005)
20. Castro, M., Costa, M., Rowstron, A.: Peer-to-peer overlays: structured, unstructured, or both? In: Technical Report MSR-TR-2004-73. Microsoft Research, System and Networking Group, Cambridge (UK) (2004)
21. Clark, J.: Google Turning Its Lucrative Web Search Over to AI Machines. Bloomberg Technology. https://www.bloomberg.com/news/articles/2015-10-26/google-turning-its-lucrative-web-search-over-to-ai-machines (2015). Accessed 17 Feb 2018
22. Coltzau, H.: Dezentrale Netzwelten als Interaktions- und Handelsplattformen. PhD thesis. University of Hagen (2012)
23. Coltzau, H., Kubek, M.: User-triggered structural changes in OSN-alike distributed content networks. In: Recent Advances in Information and Communication Technology 2016: Proceedings of the 12th International Conference on Computing and Information Technology (IC2IT), pp. 7–15. Springer International Publishing, Cham (2016)
24. Cutts, M.: Oxford Guide to Plain English. Oxford University Press (2013)
25. de Saussure, F.: Cours de Linguistique Générale. Payot, Paris (1916)
26. Dean, B.: Google's 200 Ranking Factors: The Complete List. http://backlinko.com/google-ranking-factors (2016). Aaccessed 5 Jan 2018
27. Deerwester, S.: Indexing by latent semantic analysis. J. Am. Soc. Inf. Sci. **41**(6), 391–407 (1990)
28. Dice, L.R.: Measures of the amount of ecologic association between species. Ecology **26**(3), 297–302 (1945)
29. Dunning, T.: Accurate methods for the statistics of surprise and coincidence. Comput. Linguist. **19**(1), 61–74. MIT Press, Cambridge (1993)
30. Eberhardt, J.: Was ist (bibliothekarische) Sacherschliessung? Bibliotheksdienst **46**(5), 386–401 (2012)
31. Eberhardt, R., Kubek, M., Unger, H.: Why google isn't the future. Really Not. In: Unger, H., Halang, W. (Eds.), Proceedings der Autonomous Systems, : Fortschritt-Berichte VDI, Series 10: Informatik/Kommunikation, p. 2012. VDI, Düsseldorf (2015)
32. Estivill-Castro, V.: Why so many clustering algorithms: a position paper. ACM SIGKDD Explor. Newsl. **4**(1), 65–75. ACM New York, NY, USA (2002)
33. Flake, G.W., Tarjan, R.E., Tsioutsiouliklis, K.: Graph clustering and minimum cut trees. Int. Math. **1**(4), 385–408 (2003)
34. Gaber, M.M., Stahl, F., Gomes, J.B.: Pocket Data Mining—Big Data on Small Devices. Springer International Publishing (2014)
35. Gabrilovich, E., Shaul, M.: Computing semantic relatedness using Wikipedia-based explicit semantic analysis. In: Proceedings of the International Joint Conference on Artificial Intelligence, pp. 1606–1611. Morgan Kaufmann Publishers, San Francisco, CA (2007)
36. Gantert, K.: Bibliothekarisches Grundwissen, 9th edn. De Gruyter (2016)
37. Google Inc.: Google Search Quality Rating Guidelines 2017. http://static.googleusercontent.com/media/www.google.com/en//insidesearch/howsearchworks/assets/searchqualityevaluatorguidelines.pdf (2017). Accessed 5 Jan 2018
38. Google Inside Search.: Website of Google Knowledge Graph. https://www.google.com/intl/bn/insidesearch/features/search/knowledge.html (2018). Accessed 6 January 2018
39. Guo, S., Ramakrishnan, N.: Mining linguistic cues for query expansion: applications to drug interaction search. In: CIKM '09 Proceedings of the 18th ACM conference on Information and knowledge management, pp. 335–344. ACM, New York, NY, USA (2009)

40. Hassan, S., Mihalcea, R.: Semantic relatedness using salient semantic analysis. In: Proceedings of the Twenty-Fifth AAAI Conference on Artificial Intelligence AAAI'11, pp. 884–889. AAAI Press, Cambridge, MA (2011)
41. Haveliwala, T.: Topic-Sensitive PageRank. In: Proceedings of the Eleventh International World Wide Web Conference. Honolulu, Hawaii (2002)
42. Hawkins, J., Blakeslee, S.: On Intelligence. Henry Holt and Company (2004)
43. Haynes, M.: Your Google Algorithm Cheat Sheet: Panda, Penguin, and Hummingbird. https://moz.com/blog/google-algorithm-cheat-sheet-panda-penguin-hummingbird (2014). Accessed 5 Jan 2018
44. Helbig, H.: Knowledge Representation and the Semantics of Natural Language. Springer, Berlin, Heidelberg (2006)
45. Heyer, G., Holz, F., Teresniak, S.: Change of topics over time—tracking topics by their change of meaning. In: KDIR 2009: Proceedings of the International Conference on Knowledge Discovery and Information Retrieval. INSTICC Press (2009)
46. Heyer, G., Quasthoff, U., Wittig, T.: Text Mining: Wissensrohstoff Text: Konzepte, Algorithmen, Ergebnisse. W3L-Verlag, Dortmund (2006)
47. Ishii, H., Tempo, R.: Distributed Pagerank computation with link failures. In: Proceedings of the 2009 American Control Conference, pp. 1976–1981. IEEE Control Systems Society, St. Louis (2009)
48. Jaccard, P.: Étude Comparative de la Distribution Florale dans une Portion des Alpes et des Jura. Bulletin del la Société Vaudoise des Sciences Naturelles **37**, 547–579 (1901)
49. Jamieson, K.H., Cappella, J.N.: Echo Chamber: Rush Limbaugh and the Conservative Media Establishment. Oxford University Press (2008)
50. Jain, N., Dwivedi, U.: Ranking web pages based on user interaction time. In: 2015 International Conference on Advances in Computer Engineering and Applications, pp. 35–41, Ghaziabad (2015)
51. Jay, P., Shah, P., Makvana, K., Shah, P.: An approach to identify user interest by reranking personalize web. In: Proceedings of the Second International Conference on Information and Communication Technology for Competitive Strategies (ICTCS '16). ACM, New York, NY, USA (2016)
52. Kacprzyk, J., Ziolkowski, A.: Database queries with fuzzy linguistic quantifiers. IEEE Trans. Syst. Man Cybern. **16**(3), 474–479 (1986)
53. Kasneci, G., et al.: The YAGO-NAGA approach to knowledge discovery. SIGMOD Record Spec. Issue Manag. Inf. Extract. **37**(4), 41–47 (2008)
54. Kenter, T., Rijke, M. de: Short text similarity with word embeddings. In: CIKM '15 Proceedings of the 24th ACM International on Conference on Information and Knowledge Management, pp. 1411–1420. New York, NY, USA (2015)
55. Kleinberg, J.M.: Authoritative sources in a hyperlinked environment. J. ACM (JACM) **46**(5), 668–677 (1999)
56. Koster, M.: A Standard for Robot Exclusion. http://www.robotstxt.org/orig.html (1994). Accessed 5 Jan 2018
57. Kropf, P., Plaice, J., Unger, H.: Towards a web operating system. In: Proceedings of the World Conference of the WWW, Internet and Intranet (WebNet '97), pp. 994–995. Toronto, CA (1997)
58. Kubek, M.: DocAnalyser—searching with web documents. In: Autonomous Systems 2014. Fortschritt-Berichte VDI, Reihe 10 Nr. 835, pp. 221–234. VDI-Verlag Düsseldorf (2014)
59. Kubek, M.: Dezentrale, kontextbasierte Steuerung der Suche im Internet. Ph.D. thesis. University of Hagen (2012)
60. Kubek, M., Unger, H.: A concept supporting resilient, fault-tolerant and decentralised search. In: Autonomous Systems 2017. Fortschritt-Berichte VDI, Reihe 10 Nr. 857, pp. 20–31. VDI-Verlag Düsseldorf (2017)
61. Kubek, M., Unger, H.: Centroid terms as text representatives. DocEng '16. Proceedings of the 2016 ACM Symposium on Document Engineering, pp. 99–102. ACM, New York, NY, USA (2016)

62. Kubek, M., Unger, H.: Centroid terms and their use in natural language processing. In: Autonomous Systems 2016. Fortschritt-Berichte VDI, Reihe 10 Nr. 848, pp. 167–185. VDI-Verlag Düsseldorf (2016)
63. Kubek, M., Unger, H.: Towards a librarian of the web. In: Proceedings of the 2nd International Conference on Communication and Information Processing (ICCIP 2016), pp. 70–78. ACM, New York, NY, USA (2016)
64. Kubek, M., Unger, H.: On N-term Co-occurrences. In: Recent Advances in Information and Communication Technology. Advances in Intelligent Systems and Computing, vol. 265. Springer, Cham (2014)
65. Kubek, M., Unger, H.: Special-purpose text clustering. Int. J. Intell. Inf. Process. (IJIIP) **4**(4), 11–26 (2013)
66. Kubek, M., Böhme, T., Unger, H.: Spreading activation: a fast calculation method for text centroids. In: Proceedings of the 3rd International Conference on Communication and Information Processing (ICCIP 2017). ACM, New York, NY, USA (2017)
67. Kubek, M., Meesad, P., Unger, H.: User-based Document Ranking. In: ICCIP '17 Proceedings of the 3rd International Conference on Communication and Information Processing (ICCIP 2017). ACM, New York, NY, USA (2017)
68. Kubek, M., Böhme, T., Unger, H.: Empiric experiments with text-representing centroids. Lecture Notes Inf. Theory **5**(1), 23–28 (2017)
69. Kubek, M., Schweda, R., Unger, H.: Android IR—full-text search for android. Recent Advances in Information and Communication Technology 2017 (IC2IT 2017). Advances in Intelligent Systems and Computing, vol. 566, pp. 287–296. Springer, Cham (2017)
70. Kubek, M., Unger, H., Dusik, J.: Correlating words—approaches and applications. In: Computer Analysis of Images and Patterns—16th International Conference (CAIP), Proceedings, Part I, vol. 9256 of Lecture Notes in Computer Science, pp. 27–38. Springer International Publishing Switzerland, Cham (2015)
71. Kubek, M., Witschel, H.F.: Searching the web by using the knowledge in local text documents. In: Proceedings of Mallorca Workshop 2010 Autonomous Systems, Shaker, Aachen (2010)
72. Kusner, M.J., et al.: From word embeddings to document distances. In: ICML'15 Proceedings of the 32nd International Conference on International Conference on Machine Learning vol. 37, pp. 957–966. JMLR.org (2015)
73. Le, Q., Mikolov, T.: From word embeddings to document distances. In: ICML'14 Proceedings of the 31st International Conference on International Conference on Machine Learning vol. 32, pp. 1188–1196. JMLR.org (2014)
74. Lewandowski, D.: Suchmaschinen Verstehen. Springer, Berlin, Heidelberg (2015)
75. Lum, N.: The Surprising Difference Between 'Filter Bubble' and 'Echo Chamber'. https://medium.com/@nicklum/the-surprising-difference-between-filter-bubble-and-echo-chamber-b909ef2542cc (2017). Accessed 6 Jan 2018
76. Lindstrom, M.: Small Data: The Tiny Clues That Uncover Huge Trends. Hachette UK (2016)
77. MacQueen, J.B.: Some methods for classification and analysis of multivariate observations. In: Proceedings of 5th Berkeley Symposium on Mathematical Statistics and Probability, vol. 1, pp. 281–297. University of California Press (1967)
78. Manning, C.D., Raghavan, P., Schütze, H.: An Introduction to Information Retrieval. Cambridge University Press (2009)
79. McDonald, R., et al.: Non-projective dependency parsing using spanning tree algorithms. In: Byron, D., Venkataraman, A., Zhang, D. (Eds.), Proceedings of the Joint Conference on Human Language Technology and Empirical Methods in Natural Language Processing (HLT/EMNLP), pp. 523–530. ACL, Vancouver (2005)
80. Meng, W., Yu, C., Liu, K.: Building efficient and effective metasearch engines. ACM Comput. Surv. (CSUR) **34**(1), 48–89. ACM New York, NY, USA (2002)
81. Mihalcea, R., Radev, D.: Graph-Based Natural Language Processing and Information Retrieval. Cambridge University Press (2011)
82. Mikolov, T., Chen, K., Corrado, G., Dean, J.: Efficient estimation of word representations in vector space. In: Proceedings of the International Conference on Learning Representations 2013 (2013)

83. Milgram, S.: The small world problem. Psychol. Today **2**, 60–67 (1967)
84. Miller, G.A.: WordNet: a lexical database for English. Commun. ACM **38**(11), 39–41. ACM New York, NY, USA (1995)
85. Mooers, C.N.: Zatocoding applied to mechanical organization of knowledge. Am. Doc. **2**(1), 20–32 (1951)
86. Nasraoui, O., Frigui, H., Joshi, A., Krishnapuram, R.: Mining web access logs using relational competitive fuzzy clustering. In: Proceedings of the Eight International Fuzzy Systems Association World Congress (1999)
87. Negro, A.: Mining and Searching Text with Graph Databases. In: GraphAware Blog. https://graphaware.com/neo4j/2016/07/07/mining-and-searching-text-with-graph-databases.html (2016). Accessed 8 Jan 2018
88. Novet, J.: Facebook says people sent 63 billion WhatsApp messages on New Year's Eve http://venturebeat.com/2017/01/06/facebook-says-people-sent-63-billion-whatsapp-messages-on-new-years-eve/ (2017). Accessed 8 Jan 2018
89. Oppenheim, C., Smithson, D.: What is the hybrid library? J. Inf. Sci. **25**(2), 97–112 (1999). https://doi.org/10.1177/016555159902500202
90. O'Reilly, T.: What Is Web 2.0—Design Patterns and Business Models for the Next Generation of Software. http://www.oreilly.com/pub/a/web2/archive/what-is-web-20.html (2005). Accessed 6 Jan 2018
91. Page, L., Brin, S.: The anatomy of a large-scale hypertextual web search engine. Comput. Netw. ISDN Syst. **30**(1–7), 107–117. Elsevier Science Publishers B. V. (1998)
92. Page, L., Brin, S., Motwani, R., Winograd, T.: The PageRank citation ranking: bringing order to the web (Technical report). Technical Report, Stanford InfoLab, Stanford Digital Library Technologies Project. http://ilpubs.stanford.edu:8090/422/ (1999). Accessed 5 Jan 2018
93. Pagliardini, M., Gupta, P., Jaggi, M.: Unsupervised learning of sentence embeddings using compositional n-gram features. In: NAACL 2018—Conference of the North American Chapter of the Association for Computational Linguistics (2018)
94. Pariser, E.: The Filter Bubble: What The Internet Is Hiding From You. Penguin Books Limited (2011)
95. Peters, I.: Folksonomies: Indexing and Retrieval in Web 2.0. de Gruyter (2009)
96. Phinitkar, P., Sophatsathit, P.: Personalization of search profile using ant foraging approach. In: Computational Science and Its Applications—ICCSA 2010, ICCSA 2010, Lecture Notes in Computer Science, vol 6019. Springer, Berlin, Heidelberg (2010)
97. Pourebrahimi, B., Bertels, K., Vassiliadis, S.: A survey of peer-to-peer net-works. In: Proceedings of the 16th Annual Workshop on Circuits, Systems and Signal Processing (2005)
98. Preece, S.E.: A Spreading Activation Network Model for Information Retrieval. Ph.D. thesis. Champaign, IL, USA (1981)
99. Qiu, H., Hancock, E.R.: Graph matching and clustering using spectral partitions. Pattern Recogn. **39**(1), 22–34. Elsevier (2006)
100. Quasthoff, U., Wolff, C.: The poisson collocation measure and its applications. In: Second International Workshop on Computational Approaches to Collocations, IEEE, Vienna (2002)
101. Rafiei, D., Mendelzon, A.O.: What is this page known for? Computing web page reputations. In: Proceedings of the Ninth International World Wide Web Conference, pp. 823–835. Amsterdam, The Netherlands (2000)
102. Raieli, R.: Multimedia Information Retrieval: Theory and Techniques. Elsevier Science Publishers B. V. (2013)
103. Rebele, T., et al.: YAGO: a multilingual knowledge base from wikipedia, Wordnet, and Geonames. In: Proceedings 15th International Semantic Web Conference (ISWC 2016), pp. 177–185 (2016)
104. Resnik, P.: Using information content to evaluate semantic similarity in a taxonomy. In: Proceedings of the 14th International Joint Conference on Artificial Intelligence, vol. 1 (IJCAI '95), pp. 448–453. Morgan Kaufmann Publishers Inc., San Francisco, CA, USA (1995)
105. Ribino, P., et al.: A knowledge management system based on ontologies. In: 2009 International Conference on New Trends in Information and Service Science, pp. 1025–1033 (2009)

106. Riloff, E., Jones, R.: Learning dictionaries for information extraction by multi-level bootstrapping. In: Proceedings of the Sixteenth National Conference on Artificial Intelligence, pp. 474–479. Orlando (1999)
107. Rocchio, J.: Relevance feedback in information retrieval. In: The SMART Retrieval System, pp. 313–323 (1971)
108. Roget, P.: Roget's Thesaurus of English Words and Phrases. Longman, New York, NY (1987)
109. Sakarian, G., Unger, H.: Influence of decentralized algorithms on the topology evolution of distributed P2P networks. In: Proceedings of Design, Analysis and Simulation of Distributed System (DASD) 2003, pp. 12–18. Orlando, Florida (2003)
110. Salton, G., Wong, A., Yang, C.S.: A vector space model for automatic indexing. Commun. ACM **18**(11), 613–620 (1975)
111. Sankaralingam, K., Sethumadhavan, S., Browne, J.C.: Distributed Pagerank for P2P systems. In: Proceedings of the 12th IEEE International Symposium on High Performance Distributed Computing, pp. 58–68. IEEE Computer Society Press, Seattle (2003)
112. Sateli, B., Cook, G., Witte, R.: Smarter mobile apps through integrated natural language processing services. In: Mobile Web and Information Systems—10th International Conference (MobiWIS 2013), pp. 187–202. Springer (2013)
113. Sodsee, S.: Placing Files on the Nodes of Peer-to-Peer Systems. Ph.D. thesis. University of Hagen (2011)
114. Sodsee, S., Komkhao, M., Meesad, P., Unger, H.: An extended Pagerank calculation including network parameters. In: Proceedings of the Annual International Conference on Computer Science Education: Innovation and Technology (CSEIT 2010), pp. 121–126 (2010)
115. Sparrow, B., Liu, J., Wegner, D.M.: Google effects on memory: cognitive consequences of having information at our fingertips. Science **333**, 776–778 (2011)
116. Sriharee, G.: An ontology-based approach to auto-tagging articles. Vietnam J. Comput. Sci. **2**(2), 85–94. Springer, Berlin, Heidelberg (2015)
117. Sukjit, P., Kubek, M., Böhme, T., Unger, H.: PDSearch: using pictures as queries. In: Recent Advances in Information and Communication Technology, Advances in Intelligent Systems and Computing, vol. 265, pp. 255–262, Springer International Publishing (2014)
118. Taylor, J.R.: The Mental Corpus: How Language is Represented in the Mind. OUP Oxford (2012)
119. Tsai, F.S., et al.: Introduction to mobile information retrieval. IEEE Intelli. Syst. **25**(1), 11–15. IEEE Press (2010)
120. Tushabe, F., Wilkinson, M.H.: Content-based image retrieval using combined 2D attribute pattern spectra. In: Advances in Multilingual and Multimodal Information Retrieval, pp. 554–561. Springer, Berlin, Heidelberg (2008)
121. Unger, H., Wulff, M.: Cluster-building in P2P-community networks. J. Parallel Distrib. Comput. Syst. Netw. **5**(4), 172–177 (2002)
122. van Dongen, S.M.: Graph clustering by flow simulation. Ph.D. thesis. Universiteit Utrecht, Utrecht, The Netherlands (2000)
123. Varshney, L.R.: The Google effect in doctoral theses. Scientometrics **92**(3), 785–793 (2012). https://doi.org/10.1007/s11192-012-0654-4
124. W3C: OWL 2 Web Ontology Language Document Overview (Second Edition). https://www.w3.org/TR/owl2-overview/ (2012). Accessed 6 Jan 2018
125. W3C: SPARQL 1.1 Overview. https://www.w3.org/TR/sparql11-overview/ (2013). Accessed 6 Jan 2018
126. Wachsmuth, H.: Text Analysis Pipelines: Towards Ad-hoc Large-Scale Text Mining. Springer (2015)
127. Wang, X., Halang, W.A.: Discovery and selection of semantic web services. Stud. Comput. Intell. **453**. Springer, Berlin, Heidelberg (2013)
128. Weale, T., Brew, C., Fosler-Lussier, E.: Using the wiktionary graph structure for synonym detection. In: Proceedings of the 2009 Workshop on The People's Web Meets NLP: Collaboratively Constructed Semantic Resources, pp. 28–31. ACL (2009)

129. Google Search Help: Website of Google Autocomplete. https://support.google.com/websearch/answer/106230 (2018). Accessed 5 Jan 2018
130. DFG: Website of the DFG-project 'Inhaltsbasierte Suche von Textdokumenten in großen verteilten Systemen—Search for text documents in large distributed systems'. http://gepris.dfg.de/gepris/projekt/5419460 (2009). Accessed 6 Jan 2018
131. International Data Corporation: Smartphone OS Market Share, 2017 Q1. http://www.idc.com/promo/smartphone-market-share/os (2017). Accessed 8 Jan 2018
132. Website of the project Carrot2: https://project.carrot2.org/ (2018). Accessed 6 Jan 2018
133. Website of Statista: Statistiken zu WhatsApp. https://de.statista.com/themen/1995/whatsapp/ (2016). Accessed 8 Jan 2018
134. Website of Statista: Monatliches Datenvolumen des privaten Internet-Traffics in den Jahren 2014 und 2015 sowie eine Prognose bis 2020 nach Segmenten (in Petabyte). https://de.statista.com/statistik/daten/studie/152551/umfrage/prognose-zum-internet-traffic-nach-segment/ (2016). Accessed 5 Jan 2018
135. Witschel, H.F.: Global and Local Resources for Peer-to-Peer Text Retrieval. Ph.D. thesis. Leipzig University (2008)
136. Witschel, H.F.: Terminologie-Extraktion: Möglichkeiten der Kombination statistischer und musterbasierter Verfahren. Ergon-Verlag (2004)
137. Xu, J., Croft, W.B.: Query expansion using local and global document analysis. In: Proceedings of the 19th Annual International ACM/SIGIR Conference on Research and Development in Information Retrieval (SIGIR '96), pp. 4–11, Zurich (1996)
138. Yee, K., Swearingen, K., Li, K., Hearst, M.: Faceted metadata for image search and browsing. In: CHI '03 Proceedings of the SIGCHI Conference on Human Factors in Computing Systems, pp. 401–408, New York (2003)
139. Zhao, Y., Karypis, G.: Criterion functions for document clustering: Experiments and analysis. Technical report. University of Minnesota (2001)
140. Zhu, J., von der Malsburg, C.: Learning control units for invariant recognition. Neurocomputing **52–54**, 447–453 (2003)
141. Zhu, Y., Ye, S., Li, X.: Distributed PageRank computation based on iterative aggregation-disaggregation methods. In: Proceedings of the 14th ACM International Conference on Information and Knowledge Management, pp. 578–585. ACM New York, NY, USA (2005)

Printed in the United States
By Bookmasters